P9-CEY-509

SIGNS OF INTELLIGENCE

Understanding Intelligent Design

EDITED BY

WILLIAM A. DEMBSKI
and
JAMES M. KUSHINER

Brazos Press
A Division of Baker Book House Co
Grand Rapids, Michigan 49516

© 2001 by The Fellowship of St. James

Published by Brazos Press
a division of Baker Book House Company
P.O. Box 6287, Grand Rapids, MI 49516-6287

Second printing, August 2001

Printed in the United States of America

All rights reserved. No part of this publication may be reproduced, stored in a retrieval system, or transmitted in any form or by any means—for example, electronic, photocopy, recording—without the prior written permission of the publisher. The only exception is brief quotations in printed reviews.

Table 1 on page 161 was taken from *Mere Creation*, edited by William Demski, © 1998 by Christian Leadership Ministries. Used by permission of InterVarsity Press, P.O. Box 1400, Downers Grove, IL 60515.

Table 2 on page 165, the diagram of the explanatory filter on page 182, chapter 8, and chapter 13 were adapted from *Mere Creation*, edited by William Demski, © 1998 by Christian Leadership Ministries. Used by permission of InterVarsity Press, P.O. Box 1400, Downers Grove, IL 60515.

Library of Congress Cataloging-in-Publication Data

Signs of intelligence : understanding intelligent design / edited by William A. Dembski and James Kushiner.
    p.   cm.
Includes bibliographical references (p.   ).
ISBN 1-58743-004-5 (pbk.)
1. Evolution (Biology)—Religious aspects. 2. God—Proof, Teleological.
3. Religion and science. I. Dembski, William A., 1960- II. Kushiner, James.
BL263.S54   2001
213—dc21

00-067612

For current information about all releases from Brazos Press, visit our web site:
http://www.brazospress.com

CONTENTS

PREFACE

Signs of Intelligence is the result of a modest project conceived in 1998 by Bill Dembski and me while attending a conference at Cambridge University celebrating the centennial of the birth of C. S. Lewis.

When I met Bill, he was already a key player in the growing intelligent design movement. Briefly, intelligent design asks whether the nature and structure of the material universe and the life discernible therein show evidence of being intelligently designed or not. Dembski's scholarship has brought a new scientific and mathematical rigor to an old question, particularly through his recent books (*The Design Inference* and *Intelligent Design*).

Because of our common interest in the subject, Bill and I agreed at Cambridge to publish more than a dozen articles on intelligent design by various authors from different fields in a future issue of the journal that I edit, *Touchstone*, a "journal of mere Christianity." Thus came about a special "Intelligent Design" issue of *Touchstone* (July/August 1999).

Later, due to a growing demand for reprints of this issue, it became apparent that a more durable book edition was warranted, and so *Signs of Intelligence* was conceived. Now, in addition to the original articles, this book version features a new article by Bruce Gordon and a substantial new introduction by Dembski.

I wish to thank Bill for his role in bringing the authors together initially, as well as for his editing and support. I also thank the authors for their cooperation in producing this volume. In addition, I am grateful for the work of Anita Kuhn, Sam Torode, and my wife

Patricia in producing the original magazine edition. Finally, I thank Rodney Clapp, editorial director of Brazos Press, for the opportunity to publish this book, as well as Rebecca Cooper of Brazos for her patient and careful assistance.

We hope this volume will find a wide readership among laity and clergy, teachers and students. We intend not only that it educate readers in various aspects of science, scientific naturalism, and intelligent design, but also that it broaden their imagination and deepen their understanding of the world.

—James M. Kushiner
Executive Editor, *Touchstone*

WHAT INTELLIGENT DESIGN IS NOT

WILLIAM A. DEMBSKI

Not Optimal Design

Quintilian, a Latin rhetorician of the first century, offered the following advice to writers: "Write not so that you can be understood but so that you cannot be misunderstood." Quintilian's advice is especially pertinent to the growing public debate over intelligent design. This became clear to me in a recent radio interview. Skeptic Michael Shermer and paleontologist Donald Prothero were interviewing me on National Public Radio. As the discussion unfolded, I was startled to find that how they were using the phrase "intelligent design" differed sharply from how the intelligent design community uses it.

The confusion centered on what the adjective *intelligent* is doing in the phrase "intelligent design." "Intelligent" can mean nothing more than being the result of an intelligent agent, even one who acts stupidly. On the other hand, it can mean that an intelligent agent acted with skill, mastery, and *éclat*. Shermer and Prothero understood it in the latter sense, and thus presumed that intelligent design must entail optimal design. The intelligent design community, on the other hand, understands the "intelligent" in "intelligent design" simply to refer to intelligent agency (irrespective of skill or mastery) and thus separates intelligent design from optimality of design.

7

But why then place the adjective *intelligent* in front of the noun *design*? Doesn't "design" already include the idea of intelligent agency, so that juxtaposing the two becomes redundant? No, because *intelligent design* needs to be distinguished from *apparent design* on the one hand and *optimal design* on the other. Apparent design refers to something that looks designed but really isn't. Optimal design is perfect design and hence cannot exist except in some idealized realm (sometimes called a "Platonic heaven"). Unlike intelligent design, apparent and optimal design empty design of practical significance.

Consider, for instance, biology. Many biologists claim that biological systems are not actually designed and thus attempt to assimilate all biological design to either apparent or optimal design (Stephen Jay Gould, Richard Dawkins, and Francisco Ayala are masters of this strategy). This is an evasive strategy because it avoids the central question that needs to be answered, namely, the question of actual design. The automobiles that roll off the assembly lines in Detroit are intelligently designed in the sense that actual human intelligences are responsible for them. Nevertheless, even if we think Detroit manufactures the best cars in the world, it would still be wrong to say that they are optimally designed. Nor is it correct to say that they are only apparently designed.

Although attributing intelligent design to human artifacts is unobjectionable, it quickly raises eyebrows when applied to biological systems. A biological theory of intelligent design holds that a designing intelligence is required to account for the complex, information-rich structures in living systems. At the same time, it refuses to speculate about the nature of that designing intelligence. Whereas optimal design demands a perfectionistic, anal-retentive designer who has to get everything just right, intelligent design fits our ordinary experience of design, which is always conditioned by the needs of a situation and therefore always falls short of some idealized global optimum.

No real designer attempts optimality in the sense of attaining perfect design. Indeed, there is no such thing as perfect design. Real designers strive for *constrained optimization*, which is something altogether different. As Henry Petroski, an engineer and historian at Duke University, aptly remarks in *Invention by Design*: "All design involves conflicting objectives and hence compromise, and the best designs will always be those that come up with the best compromise."[1] Constrained optimization is the art of compromise between

8

conflicting objectives. This is what design is all about. To find fault with biological design—as Stephen Jay Gould regularly does—because it misses some idealized optimum is therefore gratuitous. Not knowing the objectives of the designer, Gould is in no position to say whether the designer has proposed a faulty compromise among those objectives.[2]

Nonetheless, the claim that biological design is suboptimal has been tremendously successful in shutting down discussion about design. Interestingly, that success comes not from analyzing a given biological structure and showing how a constrained optimization for constructing that structure might have been improved. This would constitute a legitimate scientific inquiry so long as the proposed improvements can be concretely implemented and do not degenerate into wish-fulfillment, where one imagines some improvement but has no idea how it can be effected or whether it might lead to deficits elsewhere. Just because we can always imagine some improvement in design doesn't mean that the structure in question wasn't designed, or that the improvement can be effected, or that the improvement, even if it could be effected, would not entail deficits elsewhere. And, of course, the charge of poor design may simply be mistaken.[3]

The success of the suboptimality objection comes not from science at all, but from shifting the terms of the discussion from science to theology. In place of *How specifically can an existing structure be improved?*, the question instead becomes *What sort of deity would create a structure like that?* Darwin, for instance, thought there was just "too much misery in the world" to accept design: "I cannot persuade myself that a beneficent and omnipotent God would have designedly created the Ichneumonidae with the express intention of their feeding within the living bodies of Caterpillars, or that a cat should play with mice."[4] Other examples he pointed to included "ants making slaves" and "the young cuckoo ejecting its foster-brother."[5] The problem of suboptimal design is thus transformed into the problem of evil. Critics who invoke the problem of evil against intelligent design have left science behind and are engaging in philosophy and theology.

Design by intelligent agency does not preclude evil. A torture chamber replete with implements of torture is designed, and the evil of its designer does nothing to undercut the torture chamber's design. The existence of design is distinct from the morality, esthetics, goodness, optimality, or perfection of design. Moreover, there are reliable indicators of design that work irrespective of whether design includes these additional features (cf. the chapters by Behe, Bradley, Meyer, and me in this volume).

Some scientists, however, prefer to conflate science and religion—and that despite being members of the National Academy of Sciences and professing that science and religion are separate and mutually exclusive realms. Consider, for instance, the following criticism of design by Stephen Jay Gould:

> If God had designed a beautiful machine to reflect his wisdom and power, surely he would not have used a collection of parts generally fashioned for other purposes. . . . Odd arrangements and funny solutions are the proof of evolution—paths that a sensible God would never tread but that a natural process, constrained by history, follows perforce.[6]

Gould is here criticizing the panda's thumb, a bony extrusion that helps the panda strip bamboo of its hard exterior and thus render the bamboo edible to the panda.

The first question that needs to be answered about the panda's thumb is whether it displays clear marks of intelligence. The design theorist is not committed to every biological structure being designed. Mutation and selection do operate in natural history to adapt organisms to their environments. Perhaps the panda's thumb is merely such an adaptation and not designed.

Even if the intelligent design of some structure has been established, it still is a separate question whether a wise, powerful, and beneficent God ought to have designed a complex, information-rich structure one way or another. For the sake of argument, let's grant that certain designed structures are not simply, as Gould puts it, "odd" or "funny," but even cruel. What of it? Philosophical theology has abundant resources for dealing with the problem of evil, maintaining a God who is both omnipotent and benevolent in the face of evil.

The line I find most convincing is that evil always "parasitizes" good. Indeed, all our words for evil presuppose a good that has been perverted. Impurity presupposes purity, unrighteousness presupposes righteousness, deviation presupposes a way (i.e., a *via*) from which we've departed, sin (the Greek *hamartia*) presupposes a target that was missed, etc. Boethius put it this way in his *Consolation of Philosophy*: "If God exists whence evil; but whence good if God does not exist?"[7]

One looks at some biological structure, and it appears evil. Did it start out evil? Was that its function when a good and all-powerful God created it? Objects invented for good purposes are regularly co-opted and used for evil purposes. Drugs that were meant to alleviate pain become sources of addiction. Knives that were meant to cut bread become implements for killing people. Political powers that were meant to maintain law and order become the means for enslaving citizens.

Within the Judeo-Christian tradition, the good that God initially intended is no longer fully in evidence. Much has been perverted. Dysteleology, the perversion of design in nature, is real. It is evident all around us. But how do we explain it? The scientific naturalist explains dysteleology by claiming that the design in nature is only apparent, that it arose through mutation and natural selection (or some other natural mechanism), and that imperfection, cruelty, and waste are to be fully expected from such mechanisms.

Nonetheless, mutation and selection are incapable of generating the highly specific, complex, information-rich structures in nature that signal not merely apparent but actual design—that is, intelligent design. Organisms display the hallmarks of intelligently engineered high-tech systems: information storage and transfer; functioning codes; sorting and delivery systems; self-regulation and feed-back loops; signal transduction circuitry; and everywhere, complex arrangements of mutually-interdependent and well-fitted parts that work in concert to perform a function. For this reason, University of Chicago molecular biologist James Shapiro, who refuses to count himself as a design theorist, regards Darwinism as almost completely unenlightening for understanding biological complexity and prefers an information processing model.[8] Design theorists take this one step further, arguing that information processing presupposes a programmer.

Intelligent design is scientifically unobjectionable. Whether it is theologically objectionable is another matter.[9] More often than we would like, design in nature has gotten perverted. But the perversion of design—dysteleology—is not explained by issuing blanket denials of design, but by accepting the reality of design and meeting the problem of evil head on. The problem of evil is a theological problem. To force a resolution of this problem by reducing all design in nature to apparent design is an evasion. It avoids the scientific challenge posed by intelligent design. It also avoids the hard

11

work of faith, whose task is to focus on the light of God's truth and thereby dispel evil's shadows.

Not Religiously Motivated

If the discussion until now has seemed unduly theological, it is because critics of intelligent design are preoccupied with theological concerns like the problem of evil. At the same time, critics of intelligent design charge that design theorists are preoccupied with their own theological concerns. Indeed, critics of intelligent design typically regard the opposition of design theorists to Darwinian theory as motivated not by a concern for truth but by a deep fear that Darwinism undercuts traditional morality and religious belief.[10] For such critics it is inconceivable that someone, once properly exposed to Darwin's theory, could doubt it. It is as though Darwin's theory were one of Descartes's clear and distinct ideas that immediately impels assent. Thus for design theorists to oppose Darwin's theory requires some hidden motivation, like wanting to shore up traditional morality or being a closet fundamentalist.

For the record, therefore, let's be clear that the opposition of design theorists to Darwinian theory rests in the first instance on strictly scientific grounds. Yes, we are interested in and frequently write about the theological and cultural implications of Darwinism's imminent demise and replacement by intelligent design (cf. the initial chapters in this volume). But the only reason we take seriously such implications is because we are convinced that Darwinism is on its own terms an oversold and overreaching scientific theory.

Darwinism has achieved the status of inviolable science. Consequently, in challenging Darwinian theory, design theorists encounter a ruthless dogmatism. The problem is not simply that Darwinists don't hold their theory tentatively. No scientist with a career invested in a scientific theory is going to relinquish it easily. By itself, a scientist's lack of tentativeness poses no danger to science. It only becomes a danger when it turns to dogmatism. Typically, a scientist's lack of tentativeness toward a scientific theory simply means that the scientist is convinced the theory is substantially correct. Scientists are fully entitled to such convictions. On the other hand, scientists who hold their theories dogmatically go on to assert that their theories *cannot* be incorrect. Moreover, scientists who are ruthless in their dogmatism regard their theories as inviolable and critics as morally and intellectually deficient.

How can a scientist keep from descending into dogmatism? The only way I know is to look oneself squarely in the mirror and continually affirm: *I may be wrong . . . I may be massively wrong . . . I may be hopelessly and irretrievably wrong*—and mean it! It's not enough just to mouth these words. We need to take them seriously and admit that they can apply even to our most cherished scientific beliefs (this holds as much for design theorists as for Darwinists). Human fallibility is real and can catch us in the most unexpected places.

A simple induction from past scientific failures should be enough to convince us that the only thing about which we cannot be wrong is the possibility that we might be wrong. This radical skepticism cuts much deeper than Cartesian skepticism, which always admitted some privileged domains of knowledge that were immune to doubt (for Descartes, mathematics and theology constituted such domains). At the same time, this radical skepticism is consonant with an abiding faith in human inquiry and its ability to render the world intelligible. Indeed, the conviction with which scientists hold their scientific theories, so long as it is free of dogmatism, is just another word for faith. This faith sees the scientific enterprise as fundamentally worthwhile even if any of its particular claims and theories is subject to ruin.

In place of faith in the scientific enterprise, dogmatism substitutes unreasoning certainty in particular claims and theories of science. Dogmatism is always a form of self-deception. If Socrates taught us anything, it's that we always know a lot less than we think we know. Dogmatism deceives us into thinking we have attained ultimate mastery and that divergence of opinion is futile. Self-deception is the original sin. Richard Feynman put it this way: "The first principle is that you must not fool yourself, and you are the easiest person to fool." Feynman was particularly concerned about applying this principle to the public understanding of science: "You should not fool the laymen when you're talking as a scientist. . . . I'm talking about a specific, extra type of integrity that is [more than] not lying, but bending over backwards to show how you're maybe wrong."[11]

Sadly, Feynman's sound advice almost invariably gets lost when Darwin's theory is challenged. It hardly makes for a free and open exchange of ideas when biologist Richard Dawkins asserts, "It is absolutely safe to say that if you meet somebody who claims not to believe in evolution, that person is ignorant, stupid, or insane (or wicked, but I'd rather not consider that)."[12] Nor does philosopher Michael Ruse help matters when he trumpets, "Evolution is a fact, *fact, FACT!*"[13] Nor, for that matter, does Stephen Jay Gould's protegé Michael Shermer promote insight into the Darwinian mechanism of natural selection when he announces, "No one, and I mean *no one*, working in the field is debating whether natural selection is the driving force behind evolution, much less whether evolution happened or not."[14]

Such remarks, and the overweening confidence behind them, do nothing to alleviate the ongoing controversy over Darwinian evolution. Gallup polls consistently indicate that only about 10 percent of the population of the United States accepts the sort of evolution advocated by Dawkins, Ruse, and Shermer, that is, evolution in which the driving force is the Darwinian selection mechanism. The rest of the population is committed to some form of intelligent design.[15]

Science, of course, is not decided by opinion polls. Nevertheless, the overwhelming rejection of Darwinian evolution in the population at large is worth pondering. Although Michael Shermer exaggerates when he claims that no research biologist doubts the power of natural selection, he is certainly right in claiming that this is the majority position among biologists.

Why has the biological community failed to convince the public that natural selection is the driving force behind evolution and that evolution so conceived (i.e., Darwinian evolution) can successfully account for the full diversity of life? This question is worth pondering because in most other areas of science the public prefers to sign off on the considered judgments of the scientific community (science, after all, holds considerable prestige in our culture). Why not here? Steeped as our culture is in the fundamentalist-modernist controversy, the usual answer is that religious fundamentalists, blinded by their dogmatic prejudices, willfully refuse to acknowledge the overwhelming case for Darwinian evolution.

Although there may be something to this charge, fundamentalist intransigence cannot be solely responsible for the overwhelming rejection of Darwinian evolution by the public. First, fundamentalism, in the sense of strict biblical literalism, is a minority position among religious believers. Second, most religious traditions do not make a virtue out of alienating the culture. The religious world by and large would rather live in harmony with the scientific world. Despite postmodernity's inroads, science retains tremendous cultural prestige. Further, most religious believers accept that species have undergone significant changes over the course of natural history and therefore that evolution in some sense has occurred (consider, for instance, Pope John Paul II's recent qualified endorsement of evolution). The question for religious believers and the public more generally is the extent of evolutionary change and the mechanism underlying evolutionary change—in particular, whether chance and necessity alone are sufficient to explain all of life.

I submit that the real reason the public continues to resist Darwinian evolution is because the Darwinian mechanism of chance variation and natural selection seems inadequate to account for the full diversity of life. One frequently gets the sense from reading publications by the National Academy of Sciences, the National Center for Science Education, and the National Association of Biology Teachers that the failure of the public to accept Darwinian evolution is a failure in education. If only people could be made to understand Darwin's theory properly, we are told, they would readily sign off on it.

This presumption—that the failure of Darwinism to be accepted is a failure of education—leads easily to the charge of fundamentalism once education has been tried and found ineffective. For what else could be preventing Darwinism's immediate and cheerful acceptance except religious prejudice? It seems ridiculous to convinced Darwinists that the fault might lie with their theory and that the public might be picking up on faults inherent in their theory. And yet that is exactly what is happening.

The public need feel no shame at disbelieving and openly criticizing Darwinism. Most scientific theories these days are initially published in specialized journals or monographs, and are directed toward experts assumed to possess considerable technical background. Not so with Darwin's theory. The *locus classicus* for Darwin's theory remains his *Origin of Species*. In it Darwin took his case directly to the public. Contemporary Darwinists likewise continue to take their case to the public. The books of Richard Dawkins, Daniel Dennett, Stephen Jay Gould, Edward O. Wilson, and a host of other biologists and philosophers aim to convince a skeptical public about the merits of Darwin's theory. These same authors commend the public when it finds their arguments convincing. But when the public remains unconvinced, commendation turns to condemnation. Daniel Dennett even warns parents who teach their children that man is not a product of evolution by natural selection, that "those of us who have freedom of speech will feel free to describe your teachings as the spreading of falsehoods, and will attempt to demonstrate this to your children at our earliest opportunity."[16]

How can the public be commended for its scientific acumen when it accepts Darwinian evolutionary theory, but disparaged for its scientific insensibility when it doubts that same theory? The mark of dogmatism is to reward conformity and punish dissent. If contemporary science does indeed belong to the culture of rational discourse, then it must repudiate dogmatism and authoritarianism in all guises. If the public can be trusted to evaluate the case for Darwinism—and this is what Darwinists tacitly assume whenever they publish books on Darwinism for the public—then it is unfair to turn against the public when it decides that the case for Darwinism is unconvincing.

Why does the public find the case for Darwinism unconvincing? Fundamentalism aside, the claim that the Darwinian mechanism of chance variation and natural selection can generate the full range of biological diversity strikes people as an unwarranted extrapolation from the limited changes that mechanism is known to effect in practice. The hard empirical evidence for the power of the Darwinian mechanism is in fact quite limited (e.g., finch beak variation, insects developing insecticide resistance, and development in bacteria of antibiotic resistance). For instance, finch beak size does vary according to environmental pressure. The Darwinian mechanism does operate here and accounts for the changes we observe. But that same Darwinian mechanism is also supposed to account for how finches arose in the first place. This is an extrapolation. Strict Darwinists see it as perfectly plausible. The public remains unconvinced.

But shouldn't the public simply defer to the scientists? After all, they are the experts. But which scientists? It's certainly the case that the majority of the scientific community accepts Darwinism. But science is not decided at the ballot box, and Darwinism's acceptance among scientists is hardly universal. The theory of intelligent design is quickly gaining advocates at the highest level of the academy, both in the humanities and in the sciences.

Whether intelligent design is the theory that ultimately overturns Darwinism is not the issue facing scientists. The issue is whether the scientific community is willing to eschew dogmatism and admit as a live possibility that even its most cherished views might be wrong. Scientists have been wrong in the past and will continue to be wrong,

16

both about niggling details and about broad conceptual matters. Darwinism is one scientific theory that attempts to account for the history of life, but it is not the only scientific theory that could possibly account for it. It is a widely disputed theory, one that is facing ever more trenchant criticisms and that, like any other scientific theory, needs periodic reality checks.

Neither Mechanism Nor Magic

The final misconception about intelligent design I want to clear up is that it substitutes magic for mechanism—alternatively, that it invokes a supernatural cause where an ordinary natural cause will do. This misconception pervades the work of Robert Pennock. Pennock is a philosopher with a mission—to keep science safe from intelligent design. His recent book *Tower of Babel* was published by MIT Press and was even nominated for the Pulitzer Prize.[17] In that book Pennock targeted "intelligent design creationism," as he calls it.

To understand his criticism, we need first to consider more closely how the intelligent design community understands intelligent design. Proponents of intelligent design regard it as a scientific research program that investigates the effects of intelligent causes. Note that intelligent design studies the *effects* of intelligent causes and not intelligent causes *per se*. Intelligent design does not try to get into the head of a designing intelligence; rather, it looks at what a designing intelligence does and draws inferences from there.

Intelligent design is at once old and new. It is old because many special sciences already fall under it. Forensic science, intellectual property law, cryptography, random number generation, and the SETI program (Search for Extraterrestrial Intelligence) all look at features of the world and try to infer an intelligent cause responsible for those features. Intelligent design gets controversial when one takes its methods for detecting design in human contexts and shifts them to the natural sciences where no embodied, reified, or evolved intelligence could have been present. What if, for instance, the methods of intelligent design are applied to biology and show that biological systems are in fact designed? The application of intelligent design to the natural sciences is both novel and threatening, and has prompted full-scale rebuttals like Pennock's.[18]

17

Why is intelligent design threatening to the scientific establishment? Ever since Darwin, science has assumed no divine architect was needed to start creation on its course. Consequently, any designing agents, including ourselves, must result from a long evolutionary process that itself was not designed. Designing agents like ourselves, therefore, arise at the end of an undesigned natural process, and cannot exist prior to it. But if there is design in biology and cosmology, then that design could not be the work of an evolved intelligence. Rather, it must be that of a transcendent intelligence. Enter "the big G." If there's a designer behind biology and cosmology, the options for who that designer is are quite limited, with God being the preferred option. But for God to play a substantive role in science is more than our cultural elite can handle.

Hence the increasing attacks against intelligent design. What underlies these critiques is one main worry: To permit a transcendent designer into science will destroy science, reintroducing all sorts of magical, superstitious, and occult entities that modern science has thankfully banished from our understanding of the world.

Pennock gives particularly apt expression to this worry in his criticism of Phillip Johnson. According to Pennock, Johnson's position on intelligent design raises a particularly worrisome legal consequence: Johnson advocates "that science admit the reality of supernatural influences in the daily workings of the world." But what if the same reasoning that Johnson is trying to import into science were adopted in Johnson's own area of specialization—law (Johnson is a law professor)? Here's the concern as Pennock lays it out in *Tower of Babel*:

> For the law to take [Johnson's view] seriously as well, it would have to be open to both suits and defenses based on a range of possible divine and occult interventions. Imagine the problems that would result if the courts had to accept legal theories of this sort. How would the court rule on whether to commit a purportedly insane person to a mental hospital for self-mutilation who claims that the Lord told her to pluck out her eye because it offended her? How would a judge deal with a defendant, Abe, accused of attempted murder of his son, Ike, who claims that he was only following God's command that he kill Ike to prove his faith?[19]

Implicit in this passage and throughout Pennock's book is a forced choice between mechanism and magic: Either the world works by mechanisms that obey inviolable natural laws and that admit no break in the chain of natural causation, or pandemonium breaks loose and the world admits supernatural interventions that ruin science and our understanding of the world generally (and legal studies in particular). Pennock is offering his readers mechanism; Johnson is offering them magic. Any reasonable person knows which option to choose.

But as with most forced choices, there's a *tertium quid* that Pennock has conveniently ignored, and that when properly understood shows that the real magician here is Pennock and not Johnson. The *tertium quid* is intelligent design, which is entirely separable from magic. Pennock, as a trained philosopher, knows that design is an old notion that requires neither magic nor miracles nor a creator. The ancient Stoics, for instance, had design without supernatural interventions or a transcendent deity.[20] Intelligent design is detectable; we do in fact detect it; we have reliable methods for detecting it; and its detection involves no recourse to the supernatural. As the chapters in this volume will demonstrate, design is common, rational, and objectifiable.

Design theorists have a bona fide information-theoretic criterion for detecting design. What do Pennock and his fellow evolutionary naturalists have? I submit that they have not a well-supported scientific theory but a form of magic that masquerades as a scientific theory. Indeed, the real magician in Pennock's *Tower of Babel* is not Phillip Johnson and his fellow design theorists, but rather Pennock and his fellow evolutionary naturalists.

How so? There are at least three forms of magic. One is the art of illusion, where appearances are carefully crafted to conflict with reality. As entertainment, this form of magic is entirely unobjectionable. Another form of magic is to invoke the supernatural to explain a physical event. To call this magic is certainly a recent invention, since it makes most theists into magicians (Was Thomas Aquinas a magician for accepting as a historical fact the resurrection of Jesus? Was Moses Maimonides a magician for thinking that his namesake had parted the Red Sea?). According to Pennock, intelligent design is guilty of this form of magic. Nonetheless, as a professional philosopher Pennock must realize that intelligent design can avoid this charge.

Pennock is guilty of his own form of magic, however. This third form of magic is the view that something can be gotten for nothing. This form of magic can be nuanced. The "nothing" here need not be an absolute nothing. And the transformation of nothing into something may involve minor expenditures of effort. For instance,

19

the magician may need to utter "abracadabra" or "hocus-pocus." Likewise, the Darwinian just-so stories that attempt to account for complex, information-rich biological structures are incantations that give the illusion of solving a problem but in fact merely cloak ignorance.

Darwinists, for instance, explain the human eye as having evolved from a light sensitive spot that successively became more complicated as increasing visual acuity conferred increased reproductive capacity on an organism.[21] In such a just-so story, all the historical and biological details in the eye's construction are lost. How did a spot become innervated and thereby light-sensitive? How exactly did a lens form within a pinhole camera? With respect to embryology, what developmental changes are required to go from a light-sensitive sheet to a light-sensitive cup? None of these questions receives an answer in purely Darwinian terms. Darwinian just-so stories are no more enlightening than Rudyard Kipling's original just-so stories about how the elephant got its trunk or the giraffe its neck. Such stories are entertaining, but they hardly engender profound insight.

The great appeal behind this third form of magic is the offer of a bargain—indeed an incredible bargain for which no amount of creative accounting can ever square the books. The idea of getting something for nothing is not uncommon in science. In cosmology, Alan Guth, Lee Smolin, and Peter Atkins all claim that this marvelous universe could originate from quite unmarvelous beginnings (a teaspoon of ordinary dust for Guth, black-hole formation for Smolin, and set-theoretic operations on the empty set for Atkins).[22] In biology, Jacques Monod, Richard Dawkins, and Stuart Kauffman claim that the panoply of life can be explained in terms of quite simple mechanisms (chance and necessity for Monod, cumulative selection for Dawkins, and autocatalysis for Kauffman).[23]

We have become so accustomed to this something-for-nothing way of thinking that we no longer appreciate just how deeply magical it is. Consider, for instance, the following evolutionary account of neuroanatomy by Melvin Konner, an anthropologist and neurologist at Emory University: "Neuroanatomy in many species— but especially in a brain-ridden one like ours—is the product of sloppy, opportunistic half-billion year [evolution] that has pasted together, and only partly integrated, disparate organs that evolved in different animals, in different eras, and for very different purposes."[24] And since human consciousness and intelligence are said

20

to derive from human neuroanatomy, it follows that these are themselves the product of a sloppy evolutionary process.

But think about what this means. How do we make sense of "sloppy," "pasted together," and "partly integrated," except with reference to "careful," "finely adapted," and "well integrated." To speak of hodge-podge structures presupposes that we have some concept of carefully designed structures. And of course we do. Humans have designed all sorts of engineering marvels, everything from Cray supercomputers to Gothic cathedrals. But that means, if we are to believe Melvin Konner, that a blind evolutionary process (what Richard Dawkins calls the "blind watchmaker") cobbled together human neuroanatomy, which in turn gave rise to human consciousness, which in turn produces artifacts like supercomputers, which in turn are not cobbled together at all but instead carefully designed. Out pop purpose, intelligence, and design from a process that started with no purpose, intelligence, or design. This is magic.

Of course, to say this is magic is not to say it is false. It is, after all, a logical possibility that purpose, intelligence, and design can emerge by purely mechanical means out of a physical universe initially devoid of these. Intelligence, for instance, may just be a survival tool given to us by an evolutionary process that places a premium on survival and that is itself not intelligently guided. The basic creative forces of nature might be devoid of intelligence. But if that is so, how can we know it? And if it is not so, how can we know that? It does no good simply to presuppose that purpose, intelligence, and design are emergent properties of a universe that otherwise is devoid of these.

The debate about whether nature has been front-loaded with purpose, intelligence, and design is not new. Certainly the ancient Epicureans and Stoics engaged in this debate. The Stoics argued for a design-first universe: the universe starts with design and any subsequent design results from the outworkings of that initial design (they resisted subsequent novel infusions of design). The Epicureans, on the other hand, argued for a design-last universe: the universe starts with no design and any subsequent design results from the interplay of chance and necessity.[25]

What is new, at least since the Enlightenment, is that it has become intellectually respectable to cast the design-first position as disreputable, superstitious, and irrational; and the design-last position as measured, parsimonious, and alone supremely rational. Indeed, the charge of magic is nowadays typically made against the design-first position, and not against the design-last position, as I have done here.

But why should the design-first position elicit the charge of magic? Historically in the West, design has principally been connected with Judeo-Christian theism. The God of Judaism and Christianity is said to introduce design into the world by intervening in its causal structure. But such interventions cannot be anything but miraculous. And miracles are the stuff of magic. So goes the argument. The argument is flawed because there is no necessary connection between God introducing design into the world and God intervening in the world in the sense of violating its causal structure. Theists like Richard Swinburne, for instance, argue that God front-loads design into the universe by designing the very laws of nature.[26] Paul Davies takes a similar line.[27] Restricting design to structuring the laws of nature precludes design from violating those laws and thus violating nature's causal structure.

While design easily resists the charge of magic, it's the *a priori* exclusion of design that has a much tougher time resisting it. Indeed, the design-last position is inherently magical. Consider the following remark by Harvard biologist Richard Lewontin in the *New York Review of Books*:

> We take the side of science *in spite of* the patent absurdity of some of its constructs, *in spite of* its failure to fulfill many of its extravagant promises of health and life, *in spite of* the tolerance of the scientific community for unsubstantiated just-so stories, because we have a prior commitment, a commitment to materialism. It is not that the methods and institutions of science somehow compel us to accept a material explanation of the phenomenal world, but, on the contrary, that we are forced by our *a priori* adherence to material causes to create an apparatus of investigation and a set of concepts that produce material explanations, no matter how counterintuitive, no matter how mystifying to the uninitiated.[28] *(Emphasis in the original.)*

If this isn't magic, what is?

Even so, the scientific community continues to be skeptical of design. One worry is that design will give up on science. In place of a magic that derives something from nothing, design substitutes a designer who explains everything. Magic gets you something for nothing and thus offers a bargain. Design gets you something by presupposing something unimaginably bigger and thus asks you to sell your scientific soul. At least so the story goes.

But design can be explanatory without giving away the store. Certainly this is the case for human artifacts, which are properly explained by reference to design. Nor does design explain everything: There's no reason to invoke design to explain a random inkblot, but a Dürer woodcut is something else altogether. The point of the intelligent design research program is to extend design from the realm of human artifacts to the natural sciences. The program may ultimately fail, but it is only now being tried and it is certainly worth a try. Moreover, this program has a rigorous information-theoretic underpinning.

Just as truth is not decided at the ballot box, so truth is not decided by the price one must pay for it. Bargains are all fine and good, and if you can get something for nothing, go for it. But there is an alternative tendency in science that says that you get what you pay for and that at the end of the day there has to be an accounting of the books. Some areas of science are open to bargain-hunting and some are not. Self-organizing complex systems, for instance, are a great place for scientific bargain-hunters to shop. Bénard cell convection, Belousov-Zhabotinsky reactions, and a host of other self-organizing systems offer complex organized structures apparently for free.[29] But there are other areas of science that frown upon bargain-hunting. The conservation laws of physics, for instance, allow no bargains. The big question confronting design is whether it can be gotten on the cheap or must be paid for in kind. Design theorists argue that design admits no bargains.

Pennock and his fellow evolutionary naturalists are bargain hunters. They want to explain the appearance of design in nature without admitting actual design. That's why Richard Dawkins begins *The Blind Watchmaker* with "Biology is the study of complicated things that give the appearance of having been designed for a purpose,"[30] whereupon he requires an additional three hundred and fifty pages to show why it is only an appearance of design.

Pennock and his fellow evolutionary naturalists have my very best wishes for success in their hunt for the ultimate bargain. They may even be right. But they are not guaranteed to be right. And they certainly haven't demonstrated that they are right. Indeed, they are nowhere near pulling the rabbit out of the hat.

This volume shows that the smart money is on design.

THE INTELLIGENT DESIGN MOVEMENT

Challenging the Modernist Monopoly on Science

PHILLIP E. JOHNSON

Phillip Johnson, J.D. (University of Chicago), is Jefferson Peyser Professor of Law at the University of California at Berkeley, where he has taught law for thirty years. A frequent lecturer, he is also the author of Darwin on Trial, Reason in the Balance, Defeating Darwinism by Opening Minds, Objections Sustained, *and* The Wedge of Truth *(all InterVarsity), as well as two textbooks on criminal law. He is also an elder in the Presbyterian Church (USA).*

A new movement made its public debut at a conference of scientists and philosophers held at Southern Methodist University in March 1992, following the publication of my book *Darwin on Trial.* The conference brought together as speakers some key figures, particularly Michael Behe, Stephen Meyer, William Dembski, and me. It also brought a team of influential Darwinists, headed by Michael Ruse, to the table to discuss this proposition: "Darwinism and neo-Darwinism, as generally held in our society, carry with them an *a priori* commitment to metaphysical naturalism, which is essential to making a convincing case on their behalf." As I wrote in my introduction to the first edition of the papers from that conference,

I do not think the issue was ever really confronted on this question. . . .
What the anti-Darwinists called metaphysical naturalism the Darwinists
called "science," and they insisted that for science to cease being naturalistic would be for it to cease being science. To put the matter in the simplest possible terms, the Darwinist response to the question presented was
not "No, that is wrong, because the case for Darwinism can be made without assuming a naturalistic perspective." Instead, they answered "So what?
All that you are really saying is that Darwinism is science."

That may seem a deadlock, but the amazing thing was that a
respectable academic gathering was convened to discuss so inherently subversive a proposition. I was sure that in the long run discussions of that sort would be fatal for Darwinism because they
would reveal that the theory finds its justification in philosophy, not
evidence. Biologists have legitimate authority to tell us the facts that
they observe in the field and in their laboratories. They have no
authority to tell the rest of us what metaphysical assumptions we
must adopt. Once it becomes clear that the Darwinian theory rests
upon a dogmatic philosophy rather than the weight of the evidence,
the way will be open for dissenting opinions to get a fair hearing.
In a nutshell, that is the strategy. Now that several years have passed
and a new century is here, it is time to review how the strategy has
grown and progressed, to evaluate how far we have come, and to
forecast what we expect to accomplish in the next decade. But first
I need to explain the intellectual background in more detail.

The Background

Most persons who have written about creation and evolution
have assumed that they were entering a debate over facts and evidence, and their objective accordingly has been to state in detail
what they consider to be the facts and to support their conclusions
with evidence. Darwinian evolutionary scientists assert confidently
that the Genesis account is mythology, that the earth is billions of
years old, that the first primitive living organism emerged from a
chemical soup by some combination of chance and chemical laws,
and that life thereafter evolved to its present diversity by natural
means, guided by natural selection but not by God. Theistic evolutionists defend basically the same account, adding that the evolutionary process was sustained and guided by God in some manner

that cannot be detected by scientific investigation. Biblical creationists defend the Genesis account, arguing that Darwinian evolution is bad or biased science while differing among themselves about such important details as whether the "days" of Genesis were twenty-four-hour periods or geological epochs, and whether Noah's flood was worldwide or local. The argument never goes anywhere.

The Darwinists hold the dominant position in the sense that only their position is taught in public education and promoted in the national media. Nevertheless, they are frustrated and worried that so much resistance remains, especially in North America. Scientists, educators, museum curators, and others have made determined efforts to convince the public, but public opinion polls indicate that the public isn't getting the message. Over 40 percent of Americans seem to be outright creationists, and most of the remainder say they believe in God-guided evolution. Less than 10 percent express agreement with the orthodox scientific doctrine that humans and all other living things evolved by a naturalistic process in which God played no discernible part. These figures, from recent polls, are practically unchanged from previous polls in the early 1980s. The Darwinists hold a commanding power position for the time being, but they have not convinced the masses. The situation is sufficiently precarious that in 1998 the National Academy of Sciences found it necessary to issue a guidebook on *Teaching About Evolution and the Nature of Science* (hereafter *Guidebook*) urging public school teachers to "teach evolution"—i.e., to promote the neo-Darwinian theory—regardless of local opposition.

By "teaching about evolution" the National Academy emphatically does *not* mean that the teachers should inform students candidly about why the subject is so controversial, and it especially does not want them to make students aware of the dissenting arguments (except perhaps in caricatured form, as presented by Darwinists like Carl Sagan and Stephen Jay Gould). Instead, the *Guidebook* encourages teachers to mollify the religious people with vague reassurances that "religious faith and scientific knowledge, which are both useful and important, are different,"[1] and to deny that there is any real controversy because "there is no debate within the scientific community over whether evolution has occurred."[2] To make the controversy disappear, the *Guidebook* defines evolution so broadly ("descent with modification") that it "occurs" every time a baby is born. Who can deny that babies are born, dogs are bred, or that the gene pool is constantly being modified?

This strategy of trivializing the subject might be effective if the science educators and their allies completely controlled the channels of communication, but increasing numbers of high school and college students come to the classroom already knowing that there are reasonable grounds for dissent, advocated by persons (such as the authors represented in this book) with impressive scientific and academic credentials. The best-informed students also know that prominent writers like Richard Dawkins, Carl Sagan, Edward O. Wilson, and Daniel Dennett promote atheism in the name of evolutionary science, with the apparent approval of the scientific establishment. (Other authorities like Stephen Jay Gould purport to be more friendly to "religious belief," but only on the condition that religious authorities stick to questions of moral values and defer to science on all issues of fact.) When the National Academy dodges all the tough questions with evasive platitudes, it effectively teaches independent-minded students to regard the pronouncements of science educators with no more trust than they regard political or commercial advertisements. Eventually the scientific community will pay a high price for this campaign of prevarication.

The Two Models of "Science"

The science educators don't want to be dishonest, but they don't know any other way to deal with people who are so irrational as to deny that our existence is best explained by evolution. The educators also think that they are giving as much respect to religious belief as they honestly can, and that to be more explicit on the subject would merely cause unnecessary offense and provoke emotional opposition. In consequence, they assume that an honest dialogue is impossible, so they see no alternative but to counter the opposition with tactics of intimidation, evasion, and propaganda. Similarly, dissenters from evolutionary orthodoxy are often astonished that so many scientists cannot see that there is a genuine scientific case against Darwinism, and that widespread dissent cannot be dismissed out of hand as the product of ignorance or prejudice. Why can't eminent scientists seem to grasp the obvious point that finch beak variation does not even remotely illustrate a process capable of making birds in the first place?

28

The reason for this deadlock is quite simple. In our culture there are two distinct models of the scientific enterprise, and the persuasiveness of the case for Darwinian evolution depends entirely on which model you adopt.

In the first, the *materialist* model, science is seen as based by definition upon philosophical naturalism or materialism. For present purposes naturalism and materialism amount to the same thing. The first asserts that nature is all there is, while the second adds that nature is made up of matter, i.e., the particles that physicists study, *and nothing else*. (Philosophers tend to prefer the less familiar term *physicalism*, because it avoids the ordinary-language distinction between matter and energy—energy being also a physical entity.) Whichever term is used, every event or phenomenon is conclusively presumed to have a material cause, at least after the ultimate beginning. Within this first model, to postulate a non-material cause—such as an unevolved intelligence or vital force—for any event is to depart altogether from science and enter the territory of *religion*. For scientific materialists, this is equivalent to departing from objective reality into subjective belief. What we call intelligent design in biology is by this definition inherently antithetical to science, and so there cannot conceivably be evidence for it.

The second, or *empirical* model, defines science strictly in terms of accepted procedures for testing hypotheses, such as repeatable experiments. (I use the term "empirical" here in its dictionary sense of "arising from observation or experiment," as opposed to arising by deductive reasoning from philosophical axioms.) Of course scientific materialists also employ these testing procedures, but only up to the point where materialism itself comes into question. For true empiricists, whatever is testable by scientific methods is eligible for consideration. Within science one cannot argue for supernatural creation (or anything else) on the basis of ancient traditions or mystical experiences, but one can present evidence that unintelligent material causes were not adequate to do the work of biological creation. Whether some phenomenon could have been produced by unintelligent material causes, or whether an intelligent cause must be postulated, both ideas are eligible for investigation whether the phenomenon in question is a possible prehistoric artifact, a radio signal from space, or a biological cell.

If you adopt the materialist model, a materialistic evolutionary process that is at least roughly like neo-Darwinism follows as a mat-

29

ter of deductive logic, regardless of the evidence. Otherwise, how could complex organisms exist? To say that they are the product of design by an unevolved intelligence, even one that works by guiding evolution, would be to repudiate materialism and hence to abandon science. Before life, especially intelligent life, can come into existence, it must evolve from unintelligent matter by a naturalistic mechanism that must by definition be unintelligent. That mechanism must employ some combination of random variation and physical law (the principle of natural selection being a sort of law) because nothing else could have been available.

This kind of deductive reasoning is so overpowering to materialists that Darwinists sometimes say that their theory is as self-evidently true as the basic principles of arithmetic. Evolutionary biologist Paul Ewald exemplifies this Darwinian logic:

> Darwin only had a couple of basic tenets. . . . You have heritable variation, and you've got differences in survival and reproduction among the variants. That's the beauty of it. It has to be true—it's like arithmetic. And if there is life on other planets, natural selection has to be the fundamental organizing principle there, too.[3]

The fallacy here is that from the proposition "heritable variation and differential survival occur," it does not follow that these factors have any substantial creative power.

Scientific empiricists, as I use the term, hold that there are three kinds of causes to be considered rather than only two. Besides chance and law, there is also agency, which implies intelligence. Intelligence is not an occult entity, but a familiar aspect of everyday life and scientific practice. No one denies that such common technological artifacts as computers and automobiles are the product of intelligence, nor does anyone claim that this fact removes them from the territory of science and into that of religion.

It is also common in scientific practice to infer the existence of something that is not observable (cold dark matter, extinct ancestors that were not fossilized) because it is thought necessary to explain the phenomena that *are* observable. For example, Carl Sagan's SETI (Search for Extraterrestrial Intelligence) radio telescopes search the sky for evidence of radio signals from space aliens. If they were to receive a signal containing a sequence of prime numbers, as portrayed in the movie *Contact*, they would conclude that it came from intelligent beings—without the need for independent evidence of the existence and nature of the aliens.

Evidence of intelligent design is permissible in such cases because it does not conflict with materialist metaphysics, the aliens being presumed to have evolved by natural selection. The proposition that the biological cell is the work of intelligence is out of the question for materialists not because of the evidence but because—in the words of famed Harvard University geneticist Richard Lewontin—"[our] materialism is absolute, we cannot allow a Divine Foot in the door."

The confusion between these two models sets the stage for an unproductive argument that can never go anywhere. Scientific materialists think that advocates of intelligent design (ID) are either irrational or dishonest because they are advocating as science a proposition that ought to be confined to religion, namely the claim that scientific evidence points to the reality of a designing intelligence in the origin and development of life. Moreover they claim to have positive evidence for intelligent design in biology when the rules of science-as-materialism specify that such a thing cannot exist. Materialists classify such people not as empiricists but as "creationists," a term that in materialist jargon means biblical literalism and is inherently pejorative, suggesting a combination of irrationality and intellectual dishonesty. Hence materialists insist that "creationism," including any consideration of ID, must be banned from scientific discussions, and even from public discourse altogether, as a reprehensible and unconstitutional attempt to pass off religion as science.

We who are willing to consider evidence for ID, on the other hand, think of ourselves as the true empiricists and hence the true practitioners of scientific thinking. From our standpoint it is the materialists who are the "fundamentalists," in the pejorative sense of the term, because they adhere to a metaphysical dogma in the teeth of contrary scientific evidence. If design is a legitimate subject for scientific investigation in the case of computers, communications from space aliens, and peculiar markings on cave walls, why should it be arbitrarily excluded from consideration when dealing with the biological cell or the conscious mind? Whether the evidence actually does support design hypotheses in biology is a point in dispute, of course, but in our opinion the scientific materialists effectively concede the point when they adamantly refuse to admit a distinction between "materialism" and "science." They must realize at some level that they cannot win the argument on the basis of evidence, and therefore must win it by imposing a definition of science that disqualifies their critics regardless of the evidence.

Two Examples from the National Academy of Sciences

The policy of supporting Darwinism and materialism leads science educators to present the subject in a manner that actively discourages students from cultivating the critical thinking skills that are essential in real scientific research. Students are also never prepared to understand public controversies over subjects like social Darwinism and genetic determinism because the educators present a whitewashed version of their theory. I'll give two illustrations, both involving the National Academy's *Guidebook*. I choose this particular text as an example because it is simple, recent, and has the official *imprimatur* of the nation's most prestigious scientific organization. Similar confusions abound in the literature of evolution at every level.

On page 19, the *Guidebook* describes one of the most frequently cited examples of natural selection in a section titled "Ongoing Evolution Among Darwin's Finches." Here is the complete text:

> A particularly interesting example of contemporary evolution involves the 13 species of finches studied by Darwin on the Galapagos Islands, now known as Darwin's finches. A research group led by Peter and Rosemary Grant of Princeton University has shown that a single year of drought on the islands can drive evolutionary changes in the finches. Drought diminishes supplies of easily cracked nuts but permits the survival of plants that produce larger, tougher nuts. Drought thus favors birds with strong, wide beaks that can break these tougher seeds, producing populations of birds with these traits. The Grants have estimated that if droughts occur about once every 10 years on the islands, a new species of finch might arise in only about 200 years.

A good science teacher might employ humor to illustrate the fallacy of extrapolation here. "If the average length of finch beaks in a population increases 5 percent following drought years, and droughts occur every ten years, how long will it take the beaks to grow from an average of one inch in length to ten feet, or for finches to become eagles?" It is no wonder that the *Guidebook's* authors did not quote the title of the Grant's 1987 paper in *Nature*, "Oscillating Selection in Darwin's Finches," because that would have signaled to teachers, and perhaps also to bright students, that the finch beak example involves no continuing directional change at all. The drought year in question was followed a few years later by floods, and the average beak size promptly went back to normal. But even if finches did grow steadily larger for a time, would this show that they can change into something completely different?

This example is not taken out of context, nor is it atypical. It follows the thesis of *The Beak of the Finch*[4] by Jonathan Weiner, a book that won the Pulitzer Prize in 1995 and has been enthusiastically recommended to the public by leading authorities, including the president of the National Academy of Sciences. It is easy to see why the Darwinists feel they have to present evidence in a selective and slanted manner. Under any kind of objective analysis, it would become apparent that the Darwinists have never discovered a mechanism capable of creating new complex organs or changing one kind of body plan into another. (The finch beak example is given top billing in the textbooks precisely because the other known examples of observed natural selection are even less impressive.) The Darwinist educators are determined to persuade rather than to educate, and so their textbooks have to bluff.

If a stock promoter drafted a prospectus the way the *Guidebook* presents the finch beak story, by padding assets and concealing liabilities, purchasers would be entitled to recover damages for fraud and the promoter might go to jail. Yet scientific materialists do not consider such presentations to be dishonest for the same reason that they do not consider it dishonest to omit from the high school textbooks (as they do) any mention of the sudden and mysterious appearance of the animal phyla in the Cambrian explosion (see chapter 11). Specific evidentiary problems can't be all that serious, they reason, since some materialist process has to have done all the creating regardless of the evidence. If the mechanism that produced the Cambrian explosion is not yet fully understood, this is a problem for advanced researchers. Students can't be taught everything at once, and to avoid encouraging them in unsound ways of thinking it is best not to make them aware of the kind of evidence that causes people to form doubts.

I could give many other examples of how Darwinian educational materials present scientific evidence selectively or misleadingly, but for my second example I would rather discuss an important sin of omission. Readers today are virtually assaulted with books by eminent scientific authorities presenting a materialist and determinist worldview in the name of science. The Harvard zoologist Edward

The Peppered Moth Story

The National Academy's *Guidebook* ignores the standard text-book example of evolution by natural selection, the peppered moths of the English midland forests. This moth population was predominantly light-colored in the early nineteenth century, and then became predominantly dark during the late nineteenth century. According to the textbook story, the moths rest during the day on tree trunks and are eaten there by birds. While the tree trunks were light-colored the light moths were better camouflaged, but the dark moths had the advantage after the trunks became dark due to the effects of industrial pollution. The light moths made a comeback after the advent of air pollution control laws in the 1950s.

Even taken at face value, the moth story (like the finch beak story) involves no innovation or directional change. Discoveries in the 1980s showed, moreover, that the moths do not normally rest on tree trunks. All textbook photographs of peppered moths on tree trunks were produced either by manually positioning live moths (which are torpid during the day) or by gluing dead moths to tree trunks. The textbook story is now thoroughly discredited, and its continued use shows how desperate Darwinists are to provide confirmation for their cherished theory.[5]

Wilson's 1998 book[6] *Consilience* argues that not only scientists but also theologians and literary scholars should base their work strictly on Darwinian assumptions. Philosopher Daniel Dennett describes Darwin's theory as a "universal acid; it eats through just about every traditional concept and leaves in its wake a revolutionized world view." (The view that God is a valid source of moral standards is one of those traditional concepts that Darwin's theory eats through, notwithstanding the vague reassurances science educators provide for religious parents.)

Influential evolutionary psychologists like Steven Pinker and Robert Wright describe human behavior as the product of genetic programs honed by natural selection, while eminent evolutionists of the political left, such as Stephen Jay Gould and Richard Lewontin, describe evolutionary psychology as a pseudoscience honed by prejudice. Molecular geneticists propose projects to alter the human genome, at first to eliminate specific genetic defects and then eventually to improve the species overall. They see no reason to respect the existing design of an organism that was produced by unintelligent mechanisms that could hardly be expected to do the job right.

Behind all the specific controversies lies one important question that the educators systematically evade: Are evolutionary and materialist assumptions merely a convention of scientific investigation, or are they valid for all purposes? When science educators are trying to justify excluding nonmaterialistic thoughts from the science curriculum, they tend to portray science as merely "one way of knowing," with the implication that other ways of knowing are equally valid. When you press them to specify *which* other ways of knowing are as valid as science, they can't think of any examples. It turns out that what they really mean is that science is the *only* way of knowing, and outside of science there are only subjective beliefs and feelings. A typical comment is that one can "feel" a sense of awe or beauty towards some object like the rainbow, even though we know through scientific investigation how the color spectrum is produced. Religious "faith," aesthetic "feelings," and moral "beliefs" are continually contrasted with scientific "knowledge," a division that assumes that only science provides truths that are valid for everybody.

For those who think that science is the only path to knowledge, and there are many such in the National Academy, it is important to extend the realm of science as far as possible to avoid a complete relativism on all subjects involving any question of value. This explains why pseudoscientific fads such as behaviorism, Freudianism, Marxism, and social Darwinism tend to gain so much influence, and to reappear in new guises every time they are discredited. It also explains why thinkers who don't claim scientific authority tend to teach that all knowledge is relative to particular interpretive communities. When only science is deemed capable of creating knowledge, ambitious worldview-proclaimers will either style themselves as scientists, or say that their nihilism is itself an inevitable consequence of scientific knowledge. Is it true that science is the only way of gaining objective knowledge, and that outside of science there is only subjective faith and belief? That is the message the

National Academy apparently wishes to convey, but it does so by persistent insinuation rather than explicit statement in order to maintain the pose of neutrality towards "religious belief."

The Right Question

In short, our scientific leadership is in a philosophical muddle and is only making things worse with its campaign of intimidation, factual misrepresentation, and semantic legerdemain. To put things on a more rational basis, the first thing that has to be done is to get the Bible out of the discussion. Too many people, including journalists, have seen the movie *Inherit the Wind* and have become convinced that everyone who questions Darwinism must want to remove the microscopes and textbooks from the biology classrooms and just read the book of Genesis to the students. It is vital not to give any encouragement to this prejudice, and to keep the discussion strictly on the scientific evidence and the philosophical assumptions. This is not to say that the biblical issues are unimportant; the point is rather that the time to address them will be after we have separated materialist prejudice from scientific fact.

The question for now is not whether the vast claims of Darwinian evolution conflict with Genesis, but whether they conflict with the evidence of biology. To make that question visible, it is necessary to distinguish between the dictates of materialist philosophy and the inferences that one might legitimately draw from the evidence in the absence of a materialist bias. So I put this simple question to the Darwinian establishment: What should we do if empirical evidence and materialist philosophy are going in different directions? Suppose, for example, that the evidence suggests that intelligent causes were involved in biological creation. Should we follow the evidence or the philosophy?

Scientific materialists find that question impossible to answer, or even to comprehend, because they identify materialism not only with science but also with rationality itself. In their minds the only alternative to materialism is a chaotic animism in which science is impossible because all events occur at the whim of capricious spirits, a world in which every question about causation can be answered with a shrug and the remark "it must be the will of God." This is nonsense, of course. The very idea of natural laws stems from the

36

concept that the world is ruled by a rational lawgiver, just as it is a historical fact that modern science grew out of a worldview guided by biblical theism. One of the absurdities of materialism is that it assumes that the world can be rationally comprehensible only if it is entirely the product of irrational, unguided mechanisms. Another absurdity is that the scientific mind itself was designed by natural selection, a force that rewards only superiority at reproduction and by whose standards the mind of the cockroach is every bit as effective as the mind of Einstein. On the contrary, the rationality and reliability of the scientific mind rests on the fact that the mind was designed in the image of the mind of the Creator, who made both the laws and our capacity to understand them.

Diehard materialists will never agree that there can be a contradiction between the findings of empirical science and the dictates of materialist philosophy, but more open-minded thinkers will grasp the possibility at once. To get the necessary reconsideration going, the first priority for critics of scientific materialism is to state the critique of materialism and naturalism in language that the intellectual community can recognize as legitimate. In the world of the university it is not legitimate to set up the Bible as an authority against the evidence of scientific observation, but it is very legitimate to show that people who claim to be basing their ideology on observation or neutral reasoning are actually proceeding on the basis of powerful hidden assumptions. It is also legitimate to show that a specific scientific observation—such as the finch beak example—appears to be evidence that natural selection has creative power only if you interpret the evidence with a powerful materialist bias.

The Strategy

This is where the ID movement's strategy comes in. To get the intellectual world discussing a new and possibly unwelcome question, it is not enough just to write a book or make an argument. We have to inspire a lot of people to start doing intellectual work based on the right questions, work of such high quality and persuasive force that the world cannot avoid discussing it. These thinkers have to produce books and articles that explore in detail what happens when you call materialism into question rather than take it for granted. As the discussion proceeds, the intellectual world will become gradually accus-

tomed to treating materialism and naturalism as subjects to be ana-
lyzed and debated, rather than as tacit foundational assumptions that
can never be criticized. Eventually the answer to our prime question
will become too obvious to be in doubt. When the philosophy conflicts with the evidence, real scientists follow the evidence. It will be
equally obvious that thinkers outside of science should not allow scientists to abuse their proper authority by forcing dubious philosophical assumptions on the rest of the world. The answers will take care
of themselves once the discussion is directed to the right questions.

The metaphor of a wedge portrays the modernist scientific and
intellectual world, with its materialist assumptions, as a thick and
seemingly impenetrable log. Such a log can be split wide open, however, if you can find a crack and pound the sharp edge of a wedge
into it. There are a number of inviting cracks in modernism, but
probably the most important one involves the huge gap between
the materialist and empiricist definitions of science. My own writing and speaking represents the sharp edge of this wedge. I make the
first penetration, seeking always only to legitimate a line of inquiry
rather than to win a debate, measuring success by the number of
significant thinkers I draw into the discussion, rather than by the
conclusions that they draw for the present.

There are some very gifted people following me into the gradually widening opening, taking the discussion to levels I could never
reach by myself. The first and most famous example is Michael Behe.
I explained in layman's terms why the Darwinian mechanism can't
do what it has to do, and Behe explained in scientific terms exactly
what that means when you understand how biology operates at the
molecular level. Behe's book *Darwin's Black Box*[7] has sold a lot of
copies and received a lot of reviews. The reviewers say what I knew
they would say: Behe's scientific description is accurate, but his thesis is unacceptable because it points to a conclusion that materialists
are determined to avoid. Of course, the reviewers tend to be philosophically naive souls who mix up the two models in their minds.
They think that sticking to the evidence means sticking to materialism regardless of the evidence. That kind of logic may satisfy those
who are highly prejudiced in favor of materialism, but it will not
work with those who are inclined to doubt.

After Behe comes William Dembski, with his remorselessly rigorous *The Design Inference*.[8] Dembski's philosophical and mathematical
reasoning is highly sophisticated, but his fundamental proposition, that

Two Key Terms

Darwinism: Living things originate through descent with modification (*descent* means descent from one or a few primitive ancestral forms; *modification* means natural selection of random variations).

Neo-Darwinism: Same as above, but with the process of modification cast in genetic terms (genetic mutations are the source of variations, and natural selection produces changes in gene frequencies).

intelligent causes can do things that unintelligent causes cannot do, and scientific investigation can tell the difference, is pure common sense. I attended a seminar on Dembski's ideas recently at a major university's philosophy department, where I saw from the reactions how common it is for clever people to deploy their mental agility in the service of obscurity. But Dembski put the concept of intelligent design on their mental maps, and eventually they will get used to it.

After Dembski come a lot more. My sense is that the battle against the Darwinian mechanism has already been won at the intellectual level, although not at the political level. When I debate Darwinists, they rarely try to defend examples like finch beak variation as showing a mechanism that can really create complex genetic information or the sort of molecular mechanisms that Behe's book describes. Instead, they shift the burden of proof to the skeptics, arguing that the mere fact that we don't have a satisfactory mechanism now doesn't necessarily mean that one will not be discovered at some time in the future. (For reasons previously explained, scientific materialists consider the promise of a materialist mechanism in the future to be equivalent to the demonstration of a mechanism in the present. If the whole system is as true as arithmetic, the missing mechanism will inevitably be discovered.) When they are on the defensive, Darwinists frequently dismiss the mechanism as a mere detail, insisting that all scientists are agreed that "evolution is a fact," even though they may disagree about exactly how it occurred. Evolution without a specific mechanism is too vague to be testable. The theory claims, for example, that an ancestral bacterium produced distant descendants as diverse as the worm and the lobster. How can one test such an ambitious claim if no details of the transformation are specified?

When the claim that large-scale evolutionary changes occur is made specific, then it becomes testable. So far the claim is failing the tests. Writers Paul Nelson and Jonathan Wells have shown this (see chapters 9 and 10) by describing the dissimilarity of supposed evolutionary cousins at the earlier embryonic stages, and by reviewing the literature describing attempts by biologists to change the direction of embryonic development by inducing mutations in the DNA. What the results show is that mutations either have no effect on the developing embryo or they have a damaging effect, leading to death or birth defects unless the developmental repair mechanisms can fix the damage. What mutations never do is to change the direction of development, as would have to happen if evolutionary transformation were to occur. To put it simply, you may believe on philosophical grounds that large-scale evolutionary transformations must have occurred, but this belief finds no support in the experimental evidence. If they did occur, no one knows how.

The Future

Persons who consider only the cultural power of evolutionary naturalism and see how thoroughly it dominates the contemporary mind may suppose that our critique of scientific materialism is a quixotic venture that can never succeed. On the contrary, I think our success is all but inevitable. In arguing that we should distinguish between objective empirical testing on the one hand and deductive reasoning from materialist philosophical assumptions on the other, we are making a point of elementary logic that is irresistible once it is understood. The only obstacle to a breakthrough is the longstanding prejudice, so deeply ingrained in educational practice, which says that materialism and science are the same thing, and that there cannot be evidence of design in biology because materialist prejudice forbids it. A prejudice like that can be protected for a while, but in the end reason always breaks through.

I measure our success in two ways. First, many thousands of high school and college students are reading our literature, and are responding very favorably. As they learn that the official textbooks present the evidence selectively, and even distort it in the manner illustrated by the finch beak example, many become highly motivated to challenge the dogmatic system that is being foisted on them. The most talented of these will be the leaders of the movement in the future. Second, the Darwinists are completely unable to meet our challenge at the intellectual level, and scarcely try. Their literature continues to promote the view that the only dissenters from Darwinism are religious fundamentalists who don't know about the overwhelming evidence that proves that "evolution has occurred." This caricature of the opposition works only with people who have never heard the dissenting arguments firsthand. With the growth of private schooling (including home schooling) and the Internet, it is no longer as easy as it was for educators to ensure that students hear only the official version of the story. Once independent-thinking young people have read the dissenting literature, they are not likely to be impressed with the evasive statements of the Darwinist establishment.

Success for our movement does not mean replacing one dogmatic system with another. Our objective is not to impose a solution, but to open the most important areas of intellectual inquiry to fresh thinking. If the fall of Darwinism inspires materialists to develop a new theory that can survive unbiased scientific testing, then so be it. If they can't do that, then the world will face the astonishing truth that the evidence of biology actually *supports* the popular belief that living organisms are the product of an intelligent creator, rather than a blind material force. When that realization sinks in, the next big project on the intellectual agenda will be to understand why so many brilliant people fooled themselves so completely for so long. Exploring that question will make the twenty-first century a very exciting time.

2

DESIGN AND THE
DISCRIMINATING
PUBLIC

Gaining a Hearing from Ordinary People

NANCY PEARCEY

Nancy Pearcey (M.A., Covenant Theological Seminary) is senior fellow at the Discovery Institute's Center for the Renewal of Science and Culture. She contributed to Of Pandas and People,[1] *a supplemental biology text advocating intelligent design, and is coauthor (with Charles Thaxton) of* The Soul of Science[2] *and (with Charles Colson) of* How Now Shall We Live?[3] *From 1991 to 1999 she was executive editor of "Breakpoint," a national radio commentary program.*

*E*volution has enormous purchase on the public imagination, and it's easy to understand why. Just peek into the average living room where toddlers everywhere are sitting wide-eyed before videos like *The Land Before Time* series. This series offers nothing less than an excursion into evolution. Colorful one-celled organisms arise in a blue-green primeval ocean, where they "change again and again," until they evolve into the endearing little dinos of the stories. It is a delightful, fairy-tale introduction into naturalistic evolution for children, and once a child's imagination is populated with bright, colorful images, it becomes ever more difficult for a parent to dislodge them and teach the child to think critically.

The Response to Imperial Darwinism

Despite the immense visibility of evolution in the culture, design is supplanting it for three reasons. First, there is a growing demand for help in answering the claims of Darwinism as it grows more pervasive and more intellectually imperialistic. Whereas a generation ago parents could simply ignore the challenge of Darwinism, today's parents cannot. Their children are getting a Darwinian message not only in the classroom but also in books and videos and other forms of entertainment. Today not only Christians but also theists of all stripes are being forced to respond to the naturalistic, mechanistic worldview implied by Darwinism in venues far from the science classroom.

We're talking here about a sizeable portion of the American population. A 1991 Gallup poll found that 46 percent of Americans still believe human beings came directly from the hand of the Creator, while another 40 percent believe God guided the process of evolution. (Only 9 percent accept the strict account of evolution by completely natural forces.) Moreover, a 1996 survey by the National Science Foundation found that fewer than half of Americans believe that "humans developed from earlier species." Clearly, there is a very large constituency in America eager for help in answering the imperialistic claims of Darwinism.

When I was executive editor for Chuck Colson's daily radio program, "BreakPoint," I found that the number of listener call-ins increased dramatically whenever we dealt with this topic on the air. The first time we ran a radio series on the subject, the number of calls tripled and quadrupled. When we ran the series again several months later, it broke call-in records again. We then ran a series popularizing *Reason in the Balance*[4] by Berkeley law professor Phillip Johnson, and again the calls spiked astonishingly. Consistently, we found that our radio listeners felt a need to get a handle on questions of Darwinism and evolution.

This is not to suggest that people merely are interested in science popularizations, which are readily available. The reason they are so hungry for good material on evolution is that they sense that it is really about much more than science. The scientific establishment portrays dissenters from Darwinism as backwoods rubes seeking to inject religion into the science classroom. But what these dissenters rightly note is that religion is already *in* the classroom. Even ordinary folks who know little about the scientific details are uncomfortably aware that Darwinism bootlegs a philosophy of naturalism that is implacably opposed to any form of theism.

And the more honest Darwinists say so. Francisco Ayala of the University of California at Irvine says natural selection "exclude[s] God as the explanation accounting for the obvious design of organisms."[5] Tufts philosopher Daniel Dennett praises Darwinism as a "universal acid"[6] that corrodes traditional spiritual and moral beliefs. And Oxford biologist Richard Dawkins says Darwin "made it possible to be an intellectually fulfilled atheist."[7]

Darwin himself made it clear that his theory had antireligious implications. Indeed, it was devised precisely to eliminate the idea of design in living things. The function of natural selection is, after all, to sift out harmful variations and preserve only beneficial ones. But if we admit God into the process, Darwin argued, then God would ensure that only "the right variations occurred . . . and natural selection would be superfluous."[8] In other words, you can have God *or* natural selection, but not both. Alternatively, given natural selection, God would be redundant. As historian Jacques Barzun writes, the central elements in Darwin's theory (i.e., random changes and the blind sifting of natural selection) were both proposed expressly to get rid of design and purpose in biology: "The sum total of the accidents of life acting upon the sum total of the accidents of variation . . . provided a completely mechanistic and material system"[9] to explain the development of living things.

This atheistic message is easily picked up by kids in the classroom. A popular high school textbook published by Prentice-Hall describes evolution as "random and undirected," working "without either plan or purpose." A textbook by Addison-Wesley says, "Darwin gave biology a sound scientific basis by attributing the diversity of life to natural causes rather than supernatural creation." American state schools are supposed to be neutral with regard to religion, but these statements are clearly antagonistic to all theistic religions.

In the words of John Wiester, chairman of the Science Education Commission of the American Scientific Affiliation, "Darwinism is naturalistic philosophy masquerading as science." And as a philosophy, its implications extend far beyond science. In a taped debate with Phillip Johnson, Cornell biologist William Provine outlines unflinchingly what Darwinism means for human values, flashing a list on an overhead projector: Consistent Darwinism implies "No life after death; No ultimate foundation for ethics; No ultimate meaning for life; No free will."[10] The only reason people still believe in such things, Provine said, is that they haven't realized the full implications of Darwinism.

On this point, his debating partner agrees wholeheartedly—though, of course, from the other side of the issue. As Johnson writes in *Reason in the Balance*, when lecturing against Darwinism on university campuses, he has found "that any discussion with modernists about the weaknesses of the theory of evolution quickly turns into a discussion of politics, particularly sexual politics." Why? Because modernists "typically fear that any discrediting of naturalistic evolution will end in women being sent to the kitchen, gays to the closet, and abortionists to jail."[11]

In other words, on both sides of the issue most people sense instinctively that there is much more at stake here than a scientific theory—that a link exists between the material order and the moral order. Though the fears Johnson encounters are certainly exaggerated, the basic intuition is right, for the question of our origin determines our destiny. It tells us who we are, why we are here, and how we should order our lives together in society. Our view of origins shapes our understanding of ethics, law, education—and yes, even sexuality. If life on earth is a product of blind, purposeless natural causes, then our own lives are cosmic accidents. There's no source of transcendent moral guidelines, no unique dignity for human life. On the other hand, if life is the product of foresight and design, then you and I were meant to be here. In God's revelation we have a solid basis for morality, purpose, and dignity.

Design and Alternative Worldviews

The second reason design is a winner is that it is a full-fledged scientific research program, not a narrowly conceived ideological position. As soon as one stakes a movement on some narrowly conceived position, there is a danger of splintering off into antagonistic groups and disagreeing over the details. For too long, opponents of naturalistic evolution have let themselves be divided and conquered over subsidiary issues like the age of the earth. The beauty of design is that it can unite everyone who opposes the broad, overarching claim of naturalism while providing a common framework for working on subsidiary issues as allies.

This is particularly important in selling design to the public, for the average person is put off by internal bickering and just wants help in meeting the larger challenge of naturalism, which has become not just an overarching philosophy but also a surrogate religion. A few years ago, Carl Sagan enchanted a huge television audience by presenting naturalism as an alternative religion in his PBS program *Cosmos*. The mere fact that he capitalized the word "Cosmos" (as religious believers capitalize "God") was a dead giveaway that he was gripped by a religious intensity. Indeed, whatever you take as the foundation of your worldview is, functionally speaking, your religion.

Sagan regarded the Cosmos as the only self-existing, eternal being: "A universe that is infinitely old requires no Creator." Whereas Christianity teaches that we are children of God, Sagan taught that we are children of *his* god—the Cosmos. "We are, in the most profound sense, children of the Cosmos," he intones, for it is the Cosmos that gave us birth and daily sustains us. He even offers a counterfeit mysticism: "Our ancestors worshiped the Sun, and they were far from foolish," for if we must worship something, "does it not make sense to revere the Sun and the stars?" Then there's Sagan's trademark phrase, "The Cosmos is all that is or ever was or ever will be" (the opening line in his book *Cosmos*, based on the television series). Anyone who attends a liturgical church recognizes that Sagan is offering a substitute for the *Gloria Patri* ("Glory be to the Father and to the Son and to the Holy Ghost. As it was in the beginning, is now, and ever shall be, world without end").

Sagan literally canonized the Cosmos, and far from repudiating this injection of religion into science, the scientific establishment richly rewarded him, even awarding him the National Academy of Science's Public Welfare Medal in 1994. Today his religion is taught everywhere in the public square—even in the books a child checks out of the public library. Among the most popular picture book characters for small children are the Berenstain Bears. In *The Berenstain Bears' Nature Guide*, we are invited to accompany the Berenstain family on a nature walk. After a few pages, we suddenly encounter in capital letters sprawled across a sunrise, glazed with light rays, those familiar words: Nature is "all that IS, or WAS, or EVER WILL BE!" It is Sagan's liturgy to the Cosmos, repackaged for tots. And to drive the point home, the authors have drawn a bear pointing directly at the reader—the impressionable young child— and saying, "Nature is you! Nature is me!"

Today Darwinian naturalism is pressed upon our imaginations long before we can think rationally and critically. It is presented everywhere as the only worldview supported by science, and it is rapidly usurping the role of religion. The design movement keeps its focus on these large and pressing questions that the public is most concerned about, while leaving the details open for further scientific investigation.

Design and Common Sense

Finally, design is a winner with the public because it is a scientific research program that actually makes sense to ordinary people. In the past, one of the most discouraging aspects of the creation/evolution controversy was the sheer number of scientific facts one had to master even to begin to make sense of the issues—genes, mutations, fossils, and how chemicals would react in a primeval "soup." It was simply too much for the average person to take in, and no matter how many facts you mastered, new findings were always turning up.

But design is not so much a set of facts as it is a way of reasoning. It is sometimes said that the scientific method is merely a codification of common sense, and that certainly is true of design. That's why some two centuries ago the English clergyman William Paley could illustrate the argument from design with simple examples: Suppose you find a watch on the beach; would you assume it was the product of the wind and the waves? Of course not; and since living things exhibit the same structure, they too must be products of an intelligent agent.

The design movement employs similar examples to massage people's intuitions, but then shows how they apply to actual research in the sciences. Physical chemist Charles Thaxton, coauthor of *The Mystery of Life's Origin*,[12] offers the illustration of finding the words "John Loves Mary" etched into a tree trunk; immediately you would recognize that this is not the product of natural forces. Likewise, since DNA is a message (and a much more complex one), it too is best explained as the product of an intelligent agent.

Michael Behe, in *Darwin's Black Box*, uses the illustration of a mousetrap to explain the concept of irreducible complexity in living things. You can't start with a wooden platform and catch a few mice, add a spring and catch a few more mice, add a hammer, and so on, each addition making the mousetrap function better. No, to even *start* catching mice, all the parts must be assembled from the outset. In the same way, many structures in living things are made of interdependent, coadapted parts that must all be present from the outset or it simply won't work (e.g., the bacterial flagellum, the whip-like outboard rotary motor that enables a bacterium to move through solution). Such structures cannot be assembled by any gradual, step-by-step Darwinian process.

In *How Now Shall We Live?* I illustrate the way creation functions as the fundamental plank in a Christian worldview across a broad range of subject areas. There we give the example of a place in the White Mountains of New England called the "Old Man in the Mountain," where Chuck Colson used to visit as a child. In the outline of the rocks one can detect, at a certain angle, what looks like the profile of an old man. It's an example of a "natural wonder" where wind and rain erosion has carved out a shape that resembles some familiar object.

By contrast, imagine you are driving through South Dakota and suddenly come upon a mountain bearing the unmistakable likenesses of four American presidents. Would anyone conclude that these shapes were the product of wind or rain erosion? Of course not. Immediately one realizes the work of artists, working with chisels and drills.

We intuitively recognize the products of design versus the products of natural forces. Mathematician William Dembski has now formalized this intuition in his book *The Design Inference*. We detect design, Dembski says, by applying an "explanatory filter" that first rules out chance and law. That is, scientists first determine if something is the product of merely random events by whether it is irregular, erratic, and unpredictable. If chance doesn't explain it, they next determine if it is the result of natural forces by whether it is regular, repeatable, and predictable. If neither of these standard explanations works—if something is irregular and unpredictable, *yet highly specified*—then it bears the marks of design. The four presidents' faces on Mt. Rushmore are irregular (not something we see happening generally as the result of erosion), yet specified (they fit a particular, preselected pattern). Applying the explanatory filter, the evidence clearly points to design.

The naturalistic scientist insists that the idea of design has no place in science. In fact, however, several branches of science already use the concept of design or intelligence and have even devised tests for detecting the work of an intelligent agent. Consider forensic science. When police find a body, their first question is, Was this death the result of natural causes, or was it foul play (an intentional act by an intelligent being)? Likewise, when archaeologists uncover an unusually shaped rock, they ask whether the shape is a result of weathering, or whether the rock is a primitive tool, deliberately chipped by some paleolithic hunter. When a cryptographer is given a page of scrambled letters, how does he determine whether it is just a random sequence or a secret code? When radio signals are detected in outer space, how do astronomers know whether it is a message from another civilization? In each case, there are straightforward tests for detecting the work of an intelligent agent.

Not just in science but throughout everyday life, we make the determination between natural and intelligent causes. In fact, we hardly give it a thought. Consider the children's game of finding shapes in the clouds; as adults, we know the shapes are just the result of wind and temperature acting on the water molecules. But what if we see "clouds" that spell out a message? In the classic film *Reunion in France,* set in Nazi-occupied Paris in the 1940s, a plucky pilot flies over the city every day and uses skywriting to spell out a single word: "COURAGE." No one would mistake the skywriting for an ordinary cloud; even though it is white and fluffy, we are quite certain that natural forces don't create words. The "explanatory filter" is simply a logical analysis of the way we reason in everyday experience.

Design and the Public Interest

Design is a concept that is simple, easy to explain, and based solidly on experience. It has tremendous popular appeal because it answers the public's most pressing concerns. What's more, it is poised to revolutionize science as dramatically as Newtonian physics did in the first scientific revolution.

Our times are not unlike the peak of the scientific revolution, when the presses poured forth a great stream of popularizations of Isaac Newton's theories. What these popularizations offered was not Newton's theories *per se* so much as a new, mechanistic worldview derived from them. As Voltaire put it at the time, no one actually *read* Newton, but everyone *talked* about him. In other words, the vast majority of people were only marginally interested in Newtonian physics for its own sake, but they were intensely interested in what it meant for a general view of the world—for human nature, ethics, religion, and the social order. Likewise, among the public today, most are not interested in mastering the details of biology, yet they are intensely interested in what Darwinism means for a general worldview.

The Darwinist establishment benefits enormously from portraying the debate about origins as a tempest in a teapot, driven by a small, marginalized group of Bible-thumpers. But the public knows intuitively that the great questions of human existence are at stake. "The fundamental and most far-reaching assumption of Darwinism is that life is the product of forces that are impersonal and purposeless—that life is a cosmic accident," says Phillip Johnson. "This is a philosophy that strikes *most* Americans as false, not just fundamentalists."[13]

At stake in this controversy is which worldview will permeate and shape our culture. Design is not an esoteric question relevant only to scientists. Design, especially as it relates to God creating the world, lies at the heart of all that Christians believe. And because Darwinian naturalists use all their cultural power to undercut design at every turn, today we're going to have to learn how to explain these worldview issues even to our toddlers.

3

PROUD OBSTACLES AND
A REASONABLE HOPE

The Apologetic Value of Intelligent Design

JAY WESLEY RICHARDS

Jay Wesley Richards, Ph.D. (philosophical theology, Princeton Theological Seminary), is director of program development at the Discovery Institute's Center for the Renewal of Science and Culture. He is a former teaching fellow at Princeton Theological Seminary, where he was editor of The Princeton Theological Review. *He has published in numerous journals and has recently completed a book on Christian apologetics with William Dembski.[1]*

*I*t is a mistake to view the theory of intelligent design (ID) as merely or even primarily a disguised apologetic for Christianity or theism. It is primarily a theory for how we may properly conclude that something—whether it be a human artifact or biological system—is designed. Design theorists claim that design is a fruitful, even necessary, concept for understanding many objects that fall under the purview of the natural sciences. They also offer an implicit critique of theories, particularly materialistic ones, that exclude design as a legitimate explanatory option.

Still, intelligent design, while not an apologetic strategy *per se,* is a valuable resource for Christian apologetics. Positively, not only can intelligent design become—by extension—an apologetic argument, but it also proposes a view of natural science compatible with the Christian doctrine of creation. Negatively, it not only provides a more empirically adequate framework for natural science than scientific materialism, but also presents a much-needed critique of this contemporary adversary of Christian belief.

A Brief History of Apologetics

Christian apologetics finds its taproot, appropriately enough, in Scripture. For example, Peter exhorts the Christians scattered among pagans in Asia Minor: "Always be prepared to give an answer *(apologia)* to everyone who asks you to give the reason for the hope that you have" (1 Peter 3:15). Many church fathers, with the exception of Tertullian, took this to mean that Christians should appropriate whatever resources were at their disposal to make the case for the Christian faith. They should, whenever appropriate, "plunder the Egyptians." Frequently, those plundered resources included then-current philosophical ideas, such as Neoplatonism and Stoicism.

In the thirteenth century, Thomas Aquinas—and three hundred years later, the Protestant scholastics—adopted a baptized Aristotelianism. This provided them with tools for advancing positive arguments for the existence of God. Thus Aquinas began his *Summa Theologica* by summarizing the "five ways" or five arguments for God's existence (which he developed more fully in his explicitly apologetic work *Summa contra Gentiles).* In the five ways, Aquinas defended God's existence in terms of the four Aristotelian causes (formal, material, efficient, and final), with the fifth way drawing on Neoplatonic themes. The fifth way is often called the teleological or design argument, which moves from the existence of order, design, and complexity in the world to the existence of a designer.

Aquinas believed that the existence of the one true God could be rationally established without recourse to specifically biblical revelation. Accordingly, in *Summa Theologica* he considered the God demonstrated by "natural theology" prior to treating the triune God of "revealed theology." This order reflects a common conviction that it is not only possible but also proper to offer some theological arguments from general or "nonparochial" premises. After all, one can't expect the pagan to accept Scripture as divine testimony if he has no reason to think there is a God in the first place. Aquinas, like most apologists, sought to build a bridge from what the unbeliever already knows or believes to belief in the one true God.

This strategy is not alien to Scripture, which claims that we can obtain some knowledge of God through what is traditionally called the "book of nature," apart from the "book of Scripture." This general knowledge suffices to make all of us accountable for our actions, even those who do not enjoy direct knowledge of biblical revelation. Hence Paul tells the Romans,

> The wrath of God is being revealed from heaven against all the godlessness and wickedness of men who suppress the truth by their wickedness, since what may be known about God is plain to them, because God has made it plain to them. For since the creation of the world God's invisible qualities—his eternal power and divine nature—have been clearly seen, being understood from what has been made, so that men are without excuse (Romans 1:18–20).

Of course, Paul does not say precisely how it is that these truths are made plain to us, but only that they are. Some theologians, like John Calvin, add that individuals have an intuitive capacity, a *sensus divinitatis*, which produces belief in God. If this is the case, then one's belief in God could be justified without having an argument for his existence. Apologetics, however, is usually concerned only with explicit arguments.

Apologetics usually takes the form of explicit arguments for the existence of God or in defense of certain theological doctrines and historical events, such as the incarnation and resurrection of Christ. Although there are probably hundreds of arguments for God's existence, most have been categorized as "cosmological," "teleological," or "ontological." Generally, cosmological arguments move from the existence of a contingent world to the existence of God as its cause. Teleological (or design) arguments move from the existence of the world's order and apparent design to the existence of an Intelligent Designer. And ontological arguments attempt to demonstrate that God, defined as the greatest possible being, entails that God exists.

Unfortunately, after the assaults of Hume, Kant, and Darwin, many Western intellectuals came to doubt the validity of these arguments. Of course, Christianity has never rested exclusively on such arguments, since we claim God has definitively revealed himself in Scripture. Still, valid theological arguments that appeal to the assumptions of unbelievers are valuable for apologetics. Happily, recent developments are breathing new life into a number of apologetic arguments, especially the cosmological and design arguments.

Limited Apologetics

If we were forced to categorize intelligent design as a traditional theistic argument, we would undoubtedly call it a teleological or design argument. But this must be qualified, since arguments for intelligent design are more modest than traditional teleological arguments. ID arguments are rarely deductive (the premises entail a conclusion), and, at least initially, they are not arguments for God's existence.

Consider for example Michael Behe's argument (in *Darwin's Black Box*) that certain biological entities exhibit what he calls "irreducible complexity" and are therefore designed. A thing is irreducibly complex only if it has a set of interconnected parts, all of which must be present and properly functioning for the thing itself to function properly. Behe argues that the bacterial flagellum, like many other biological systems, is irreducibly complex since it has components that are interdependent and thus require each other in order for the flagellum to work. Though plausible, this claim is controversial in the biological sciences because current orthodoxy mandates that every biological system evolved *cumulatively* by natural selection working on random genetic variation. Those variations that confer a survival advantage on an organism are "selected for" and dispersed throughout a population. Those that do not are "selected against." This explanation is known as neo-Darwinism.

The neo-Darwinian process is strictly cumulative. Consequently, it cannot produce complex systems whose parts are all interdependent and each necessary since the individual parts would not provide an advantage for survival until they are all in place. In short, this process cannot produce irreducibly complex systems. So, if neo-Darwinism is the enforced orthodoxy of the biological research community, it will not tolerate the existence of true irreducible complexity among biological systems. There has to be a catch. The systems can't truly be irreducibly complex. Neo-Darwinism is perforce blind to the existence of these systems.

But irreducible complexity is everywhere, not just in biology. Behe's favorite example is the mousetrap. One part of a mousetrap, such as the wooden base, isn't just a little less helpful than the whole thing. It doesn't just catch fewer mice than the entire mousetrap. Rather, without the base, spring, catch, and bar, all correctly ordered, it won't catch any mice. Thus a mousetrap, though simple in appearance, is irreducibly complex relative to its function. And we all know where such things come from: intelligent agents.

The neo-Darwinian explanation, in contrast, is impersonal and mechanistic. It excludes the causation unique to intelligent agents. A random and mechanized process cannot anticipate a future goal; it cannot construct a machine with a future function in mind. At best, it could "construct" complex things that aren't *irreducibly* complex. When we find irreducibly complex things, we rightly conclude that they are designed by intelligent agents. Similarly, Behe infers that some biological systems, like the bacterial flagellum, are intelligently designed. If neo-Darwinian dogma cannot accommodate them, so much the worse for neo-Darwinism. As a Roman Catholic Christian, Behe naturally concludes that the intelligent agent responsible is God. But he realizes that his argument alone is not a proof of the existence of God. At best, it is an argument—based on good sense and probability—that an intelligent agent is responsible for some biological realities.

Behe's argument does not entail (as in logically compel) a theological conclusion because it is consistent with other explanations. For instance, perhaps some advanced alien race planted fully constructed, reproducing organisms on a hospitable earth some time in the distant past. In that case, someone other than God would have designed these features of the biological world. Sure, it's far-fetched, but it's possible. For this reason, intelligent design arguments in biology do not normally entail theistic conclusions even if many people suspect God is lurking somewhere in the background.

The Extension of Intelligent Design

In his book *The Design Inference*, William Dembski argues convincingly that we infer design every day, both in ordinary life and in scientific disciplines such as archaeology, forensics, fraud detection, cryptography, and the Search for Extraterrestrial Intelligence (SETI). What Behe calls *irreducible complexity*, Dembski calls *specified complexity*. If something has specified complexity, it is highly improbable, has a high information content, and conforms to an independent pattern. Again, the only known causes for such things are intelligent agents, so there is no special pleading in inferring design in natural sciences like biology. Dembski argues that when we conclude that something has the property of specified complexity, we are justified in inferring that it is designed by an intelligent agent. Given the available options of chance, natural law, or design, this is the most reasonable conclusion.

So how is ID relevant to Christian apologetics? ID can be extended. We may envision its extension as a set of concentric circles, encompassing ever-larger swathes of nature within its explanatory domain. The biological entities Behe describes have a huge number of necessary physical conditions so precise that many scientists describe them as being "fine-tuned" for the existence of biological life. They range from the infinitesimal to the immense.

Many variables must be intricately calibrated to allow for the existence of living organisms that endure through time, such as the strength of the attraction of subatomic particles, of protons and electrons, of the electromagnetic force, of gravity between all matter in the universe. Coupled with the evidence of the big bang in cosmology and relativity in physics, these variables become relevant to the universe as a whole. These facts, if they are facts, leave us not with a necessarily existing, infinite, and eternal universe, but a contingent, fine-tuned, and finite one. It's not a leap to suspect that the universe has its origin in an intelligent agent that transcends the universe itself.

These physical properties amount only to necessary conditions for organic life, since it's conceivable that a universe with identical physical laws could exist without containing life. Physics and chemistry alone do not give us biology. The chemical composition of life's building blocks no more determines the sequence of DNA or the developmental details of embryology than the properties of ink and paper determine the words and rhythm of Shakespeare's sonnets. A picture begins to emerge of a universe constructed to "fit" biological beings, which themselves display exquisite marks of intelligent design. Moreover, when applied to the universe as a whole, the intelligent design argument folds into a cosmological argument, since it is reasonable to suppose that things that begin to exist have a cause for their existence. Needless to say, this is theologically suggestive.

Notice that design theorists always appeal to public evidences. That is, they argue from particular, observable features in the world rather than from specific biblical claims. One does not need to presuppose that in the beginning the God of Abraham, Isaac, and Jacob created the heavens and the earth to see that certain biological systems bear detectable marks of intelligent design, or that physical constants appear to be fine-tuned for life. Of course, the design inference, like all inferences, requires some presuppositions, but it does not require narrowly theological presuppositions. In this way, an intelligent design argument extended to the universe as a whole is similar to other arguments of natural theology, which proceed from general facts and premises rather than from strictly biblical ones.

Inference Rather than Proof

The main difference between this argument for intelligent design and much earlier natural theology and apologetics is that it is an "inference to the best explanation" rather than a deductive "proof" for God's existence. In this way it resembles a wide range of scientific arguments. While there may be some good deductive arguments for the existence of God, deductive proof is in most cases too stringent a requirement for anything outside of mathematics and logic. Besides, the apologist need not accept a standard of reasoning stronger than those of his materialist detractors. So there's nothing questionable in principle about using an inference to the best explanation in apologetics.

The skeptic may now ask impatiently: "But aren't most leading design theorists also theists?" The answer is "yes." Presumably, the skeptic's question is meant to discredit intelligent design by consigning it safely to theological studies. This sort of question is double-edged, however, since the design theorist could also ask: "Aren't most leading neo-Darwinists also materialists?" Again, the answer is "yes." But these are questions about motivation. For them to be relevant, one needs to show that the evidence for these positions, whether for intelligent design or for neo-Darwinism, requires contentious metaphysical assumptions.

It is here that we see the superiority of intelligent design over materialistic theories. The design theorist employs methods of reasoning used widely in a number of well-established disciplines. When extended to an explanation for the universe as a whole, all he need presuppose is that a theistic interpretation of nature is *possible*. But how controversial is this? Surely, only the most fanatical materialist will deny it.

Of course, some materialists—as Phillip Johnson forcefully argues—are constrained by a prior metaphysical commitment to seek only materialistic explanations of nature, rather than to follow the evidence where it leads. Even honest materialists are beginning to admit this, as Harvard University geneticist Richard Lewontin recently did in the *New York Review of Books*:

> Our willingness to accept scientific claims that are against common sense is the key to an understanding of the real struggle between science and the supernatural. We take the side of science *in spite of* its failure to fulfill many of its extravagant promises of health and life, *in spite of* the tolerance of the scientific community of unsubstantiated just-so stories, because we have a prior commitment to materialism. It is not that the methods and institutions of science somehow compel us to accept a material explanation of the phenomenal world, but on the contrary, that we are forced by our *a priori* adherence to material causes to create an apparatus of investigation and a set of concepts that produce material explanations, no matter how counterintuitive, no matter how mystifying to the uninitiated. Moreover, that Materialism is absolute, for we cannot allow a Divine Foot in the door.[2]

Clearly, intelligent design theory is more open to the possibilities inherent in nature than is Lewontin's materialism, since ID allows more types of explanations. After all, if intelligent design is at least possible, how rational or "scientific" is it to adopt a methodology that is blind to all possible evidence for it? Consequently, an extended intelligent design argument may be an effective apologetic for anyone willing to entertain at least the possibility that theism is true. The fact that die-hard materialists resist it is unfortunate but beside the point. The value of an apologetic argument should be judged not by the criteria of the closed-minded skeptic but by the unbeliever who is an honest seeker of truth.

Intelligent design theory also offers a natural science that is compatible with the Christian doctrine of creation, even if it neither entails nor presupposes it. In fact, for one who already believes that the universe is created, something like design theory is what one would expect. In this way, evidence of intelligent design confirms the doctrine of creation, even if it does not establish it. Thus, in its widest application, ID provides the unbeliever with some positive reasons for believing in God. As a resource, it is even more fecund than were Neoplatonism and Aristotelianism in previous eras.

Still, design theory's greatest apologetic value may be its potential to defeat the biggest stumbling block to faith in the contemporary world, namely, scientific materialism. For Christians, the most devastating consequence of materialism is its tendency to harden the minds and hearts of modern people to the gospel. If we do not defeat the root cause of our contemporary spiritual malaise, we will merely be treating symptoms. As the Princeton theologian J. Gresham Machen put it:

> False ideas are the greatest obstacles to the gospel. We may preach with all the fervor of a reformer and yet succeed only in winning a straggler here and there, if we permit the whole collective thought of the nation or of the world to be controlled by ideas which, by the resistless force of logic, prevent Christianity from being regarded as anything more than a harmless delusion.[3]

For more than a century we have heard that scientific progress has made Christian belief obsolete. Given the cultural prestige of science, this claim has prevented many from considering the Christian faith. If intelligent design theory exposes the inadequacy of materialistic explanations in the natural sciences, it will deflate this assertion, and could contribute to a renewal of Christian belief in the twenty-first century. This would be its most significant apologetic contribution.

4

THE REGENERATION
OF SCIENCE
AND CULTURE

*The Cultural Implications of Scientific Materialism
Versus Intelligent Design*

JOHN G. WEST JR.

*John G. West Jr., Ph.D. (government, The Claremont Graduate School),
is assistant professor of political science at Seattle Pacific University, senior fellow of the Discovery Institute, and associate director of the Discovery Institute's Center for the Renewal of Science and Culture. He is the
author of* The Politics of Revelation and Reason[1] *and coeditor of*
The Encyclopedia of Religion in American Politics.[2]

*T*o appreciate fully the cultural implications of intelligent
design, one first must understand the cultural damage inflicted by
scientific materialism, the paradigm it seeks to replace.

Scientific materialism claims that human beings can be wholly
explained as the material products of biology, chemistry, and environment. While the roots of materialism reach back to ancient
Greece, materialism became enshrined as the reigning philosophy
of Western culture largely due to the work of Charles Darwin. Dar-

60

win made materialism credible by explaining how man and his moral beliefs could have developed through an unplanned process of natural selection. Scientists and thinkers in other fields soon drew out the implications of Darwin's biology. In Germany, Karl Marx contended that one's ideas are the product of economic conditions and that the class struggle was the manifestation of Darwin's theory in human society. In America, behavioral psychologists like John Watson argued that the soul was a myth and reason was merely the physical processes of the brain.

Scientific materialism was dubious science and even shakier philosophy, but it had far-reaching consequences for Western society. By claiming that all human thoughts and actions are dictated by either biology or environment, scientific materialists undermined traditional theories of human freedom and responsibility. By asserting that our moral beliefs were merely the products of heredity or environment, scientific materialists laid the groundwork for moral relativism. By arguing that man should take control of the material processes that produced him to remake society, scientific materialists promoted a virulently coercive strain of utopianism.

The ideology of scientific materialism infected nearly every area of our culture, from politics and economics to literature and the arts, and many of its ideas still exert a powerful influence in the public square. Indeed, the extent of the impact of scientific materialism on modern culture is hard to overstate.

The Cultural Impact of Scientific Materialism

In economics and welfare, scientific materialism spawned the twin legacies of social Darwinism and socialism. Most people are somewhat familiar with the former philosophy. Herbert Spencer, William Graham Sumner, and other theorists applied the doctrine of "survival of the fittest" to human society, and in the process justified even the worst excesses of nineteenth-century capitalism. Fewer people recognize the link between scientific materialism and various strands of socialism. Marx wrote to Engels that Darwin's *Origin of Species* "contains the basis in natural history for our view," and socialist thought drew heavily on modern science to support its theory of economic determinism. Ambivalent socialist Jack London, author of such works as *The Call of the Wild*,[3] aptly summarized the socialist creed as being founded on that forbidding doctrine, the materialistic conception of history. Men are not the masters of their souls. They are the puppets of great, blind forces. The lives they live and the deaths they die are compulsory. All social codes are but the reflexes of existing economic conditions, plus certain survivals of past economic conditions. The institutions men build they are compelled to build. Economic laws determine at any given time what these institutions shall be, how long they shall operate, and by what they shall be replaced.

Although such purebred socialism never had a wide following in America, the economic determinism underlying it helped create the modern welfare state. For much of American history, welfare programs had focused on ministering to a person's whole being. Rather than assuming that all poverty was caused by impersonal economic forces over which individuals had no control, these earlier programs recognized that the choices people made could increase the likelihood of their being in poverty. As a result, these early welfare programs stressed the duty welfare recipients had to help solve their own problems. Rather than viewing the poor as the passive victims of heredity and environment, early welfare workers regarded them as responsible individuals who were capable of facilitating their own economic success. In the words of Washington Gladden, a pioneering minister of the social-gospel movement:

> Heredity is no excuse. . . . Your heredity is from God. He is your Father. Deeper than all other strains of ancestral tendency is this fact that your nature comes from God. . . . Environment is no excuse for you. . . . God is the great first fact in all our environment, no matter where you may be. There is no place of temptation in which he is not nearer to you than any human influence can be.[4]

As the ideology of scientific materialism spread, however, the belief in the ability of the poor to surmount their material conditions was gradually lost. As Marvin Olasky points out in *The Tragedy of American Compassion*,[5] by the 1930s and 1940s most social scientists viewed social problems as almost wholly the function of material causes. Thereafter government and private welfare programs increasingly treated poverty as a purely material question to be solved by purely material mechanisms. If someone lacked material goods, the comprehensive solution was to supply him with more material goods. The spiritual and moral aspects of poverty were ignored.

In the criminal justice system, scientific materialism likewise undermined theories of personal responsibility. In the traditional model of criminal justice, criminals were regarded as morally accountable for their actions, and their punishment reflected the just deserts for their crimes. In the new view, accountability and punishment were replaced with a model that regarded criminals as the helpless victims of environment and heredity. An early proponent of the new view was German Darwinist Ludwig Büchner, one of the nineteenth century's most popular evangelists of scientific materialism. Writing in his book *Force and Matter*,[6] Büchner claimed that modern science had proven that many criminals were "doomed or predestined to crime by a faulty or imperfect organization of mind and body."[7] As a result, he wrote, "many criminals are simply unfortunates, afflicted with insanity, partly in an incipient and partly in an acute state of development. 'Hence,' says G. Forster, 'we should do best in neither judging nor condemning anyone.'"[8]

By the 1930s Büchner's view had turned into the conventional wisdom. In one criminal justice text from the period, an American criminologist baldly declared that

> the grotesque notion of a private *entity*, spirit, soul, will, conscience or consciousness, interfering with the orderly processes of body mechanisms is invalidated by physiological and neurological science. . . . Man is no more 'responsible' for becoming wilful [*sic*] and committing a crime than the flower for becoming red and fragrant. In both instances the end products are predetermined by the nature of protoplasm and the chance of circumstances.[9]

Legal movements arising in the 1960s to expand the insanity defense, to decrease imprisonment even for violent offenders, and to decriminalize such behaviors as public drunkenness all had their roots in the ideology of scientific materialism.

Scientific materialism also reshaped the field of medicine, producing a vigorous eugenics movement designed to cleanse American society of persons regarded as somehow unfit, from the blind and deaf to prostitutes and alcoholics. By the early 1930s, thirty states had enacted forced sterilization laws—laws that ultimately served as models for the sterilization laws enacted by Nazi Germany. At the legal level, the American eugenics crusade culminated in the infamous Supreme Court decision in *Buck v. Bell*, where Justice Oliver Wendell Holmes declared that compulsory sterilization for the mentally handicapped was constitutional because "three generations of imbeciles are enough."

While the Nazi Holocaust eventually dampened American enthusiasm for eugenics, the underlying rationale has been resurrected in recent decades in abortions for sex selection, infanticide for children with disabilities, and the growing practice of assisted suicide. If humans are purely material beings, then their worth inevitably degrades as their physical capacities deteriorate. Evolutionary psychologist Peter Singer, now a professor of bioethics at Princeton University, has gone so far as to argue that human beings with severe physical disabilities should no longer be viewed as worthy of life. In his words, "if we compare a severely defective human infant with a nonhuman animal, a dog or a pig . . . we will often find the nonhuman to have superior capacities. . . . Only the fact that the defective infant is a members of the species *Homo sapiens* leads it to be treated differently from the dog or pig. Species membership alone, however, is not morally relevant."[10]

Traditional beliefs about family life and marriage have also suffered under the materialist onslaught. In such books as *The Moral Animal*[11] by Robert Wright, evolutionary psychologists have argued that adultery is programmed into men by natural selection for reproductive success. The corrosive effect on public discourse of this sort of reductionism could be seen in the discussions of President Clinton's extramarital escapades. Numerous articles in the popular press presented the evolutionary argument for adultery as a way of explaining away the President's flouting of traditional norms.

But adultery isn't the only behavior that evolutionary psychologists have sought to explain away. Steven Pinker of MIT claims that infanticide is programmed by evolution into mothers who kill their newborns. While evolutionary psychologists explain with apparent ease the functionality of some of the most horrendous behaviors known to civilized man, they become much less convincing when they try to account for the existence of positive traits, such as altruism. As biologist Jeffery Schloss has pointed out, most evolutionary psychologists cannot accept the idea that truly self-denying behavior could be produced by natural selection. So they must interpret what appears to be altruism as merely another form of selfishness. In this scheme of things, even the behavior of Mother Teresa becomes an expression of self-centeredness.

Scientific materialism has also become pervasive in the field of education. Here one of the major impacts has been the narrowing of the acceptable range of debate in the natural and social sciences. It has to be one of the great ironies of modernity that in an age when free inquiry is trumpeted as one of our culture's few remaining universals, the freedom to pursue fundamental questions in the scientific realm is routinely denied or even ridiculed. Richard Dawkins, for example, asserts that anyone who does not believe in Darwinian evolution is "ignorant, stupid, or insane." With views like these among the academic elites, is it any wonder that ideas challenging Darwin and his progeny are often ruled out from the start?

Extreme forms of multiculturalism and feminism endemic on college campuses represent further variations of scientific materialism. Proponents of these ideologies take their materialism wholesale, asserting that human beings are so dominated by their race or gender that their political views, their morality, and their religious beliefs are merely mechanical by-products.

Scientific materialism even has links to intellectual movements within academia that might at first seem opposed to it, such as postmodernism. While on the surface postmodernism seems to reject the artificially truncated rationalism embodied by scientific materialism, it actually adopts the materialist account of man as its starting point. Richard Rorty, dean of American postmodernists, even argues for the importance of "keeping faith with Darwin." It is precisely because the materialist account of man is so bleak that it creates fertile ground for postmodernism to grow. In Nietzsche's insightful phrase, Darwin's teaching is "true but deadly." If human beings (and their beliefs) really are the mindless products of their material existence, then everything that gives meaning to human life—religion, morality, beauty— is revealed to be without objective basis. To avoid sliding into nihilism as a result of this revelation, postmodernists take a page from Nietzsche and reject reason altogether, urging people to fashion their own reality through an act of the will. In this way, the narrow "rationalism" of scientific materialism begets utter irrationalism.

The Cultural Implications of Intelligent Design

Thus have the ideas of scientific materialism influenced virtually every area of modern culture. Ideas have consequences, and in the modern era scientific ideas have had particularly momentous consequences for society. That is one reason the emerging debate over intelligent design is so intriguing. If scientists can be persuaded to change fundamentally how they view man and the universe, their conversion will likely have consequences far beyond the sciences themselves. Just as scientific materialism had implications for the furthest reaches of our culture, so too intelligent design will raise an array of implications for all the disciplines that took their founding assumptions from scientific materialism. Because intelligent design is a movement in its infancy, the precise contours of its implications for culture are still emerging. But five general implications seem clear.

First, intelligent design can help reinvigorate the case for free will and personal responsibility. Modern science has reduced human action to a product of genes and environment by insisting that only material causes are relevant. In the words of a recent psychology text: "What is left besides genetics and environment?" Intelligent design, however, suggests that mind precedes matter and that intelligence is an irreducible property just like matter. This opens the door to an effective alternative to materialistic reductionism. If intelligence itself is an irreducible property, then it is improper to try to reduce mind to matter. Mind can only be explained in terms of itself—like matter is explained in terms of itself. In short, intelligent design opens the door to a theory of a nonmaterial soul that can be defended within the bounds of science. At the very least, if intelligence is understood as an irreducible property of human beings, the grounds on which science can undercut free will and personal responsibility will be significantly diminished.

What effects might this have for public policy? Consider the possibilities in just one area—welfare policy. As long as human beings are regarded as the helpless victims of material forces, it is understandable why welfare policies would focus narrowly on changing material inputs rather than also on looking at issues of character and accountability. Once men and women are regarded again as accountable beings who make real choices, however, this model becomes untenable. Welfare programs could no longer simply assume that supplying material benefits to someone will solve his problems. Instead, welfare programs would have to examine the role of a person's own choices and could hold him responsible for making good choices. Driven by the failures of a welfare system based on materialist premises, recent efforts at welfare reform have tried to do precisely this. Unfortunately, the underlying social science framework does not yet support the renewed push toward personal responsi-

66

bility in the welfare system, and because of this it remains to be seen whether the current reforms can be sustained. The long-term role of personal responsibility in welfare policy remains precarious until the scientific critique of free will has been answered convincingly. Intelligent design can help do this.

A second cultural implication of intelligent design concerns the defense of traditional morality. Modern biology seeks to understand morality primarily in terms of reproductive success, but intelligent design opens the door to understanding how morality can serve a variety of other ends. Rather than assume that all human behaviors (no matter how apparently destructive) must have been produced by natural selection and therefore promote reproductive success, intelligent design would promote honest questioning of whether certain behaviors—such as adultery—really do serve a biological function. Because intelligent design does not assume that every behavior must be tied to reproductive success, it is free to consider a range of other explanations. In the process, it may provide a powerful way to check the moral relativism spawned by scientific materialism, especially in the areas of family life and sexual behavior.

A third area where intelligent design has cultural implications is the sanctity of human life. As noted previously, the argument over eugenic abortions and assisted suicide is premised largely on the notion that human beings are the sum of their material parts. If human beings are more than this, then the cogency of this argument disintegrates. Once the idea of a nonmaterial soul gains new currency, the ethical context in which issues such as abortion and euthanasia are debated will considerably expand.

A fourth cultural implication of intelligent design is the defense of science itself. Many in the natural sciences today are disturbed at the growing backlash against scientific rationality that can be seen in both the New Age and postmodernist movements. In *Unweaving the Rainbow*,[12] outspoken materialist Richard Dawkins attacks postmodernism in particular for its rejection of scientific rationality. What materialists often fail to appreciate is that it is their truncated view of the scientific enterprise that is partly responsible for the rise of irrationalism. Human beings know intuitively that they are more than a bundle of cells and synapses. When science in the name of reason debunks this truth, is it any wonder that some people choose irrationalism? By supplying a framework for science that can account for the full richness of what human beings really are, intelligent design can remove one of the chief inspirations for the postmodern impulse and help restore the integrity of science.

A fifth cultural implication of intelligent design is its support of free inquiry. Intelligent design is more consistent with unhindered scientific exploration than scientific materialism because it admits a far wider range of possible explanations in scientific discussions. Unlike scientific materialism, which seeks to explain everything in terms of material causes, intelligent design is not monocausal. It does not deny the validity of material causation; it only denies the assertion that material causation is exhaustive. Because of this, intelligent design opens the door to the discussion of many different kinds of questions in the natural and social sciences. Instead of closing off debate, it opens it. Instead of limiting inquiry, it expands it. For many disciplines suffering under the stale hegemony of scientific materialism, intelligent design will provide a welcome invitation to rethink some of their most fundamental assumptions.

A Regenerated Science

Cultural change is not inevitable, and for intelligent design ultimately to make a difference to the rest of culture, cultural leaders must become acquainted with the emerging debate and be encouraged to apply the new paradigm to their own fields. That is one of the goals of the Discovery Institute's Center for the Renewal of Science and Culture, which not only supports primary research by those in the natural sciences, but also seeks to brief policymakers on the implications of the emerging design paradigm for their own areas. Unfortunately, many cultural leaders remain well behind the curve when it comes to recent developments in the sciences. This is just as true among conservatives as it is among liberals.

A year after the publication of Michael Behe's groundbreaking *Darwin's Black Box,* the influential conservative journal *National Review*[13] published an essay by law professor John McGinnis urging conservatives to jump on the bandwagon of modern evolutionary biology. According to McGinnis, recent discoveries during the last two decades have led to a convincing revival of Darwinism and "any political movement that hopes to be successful must come to terms with the second rise of Darwinism."

McGinnis is not alone among conservative intellectuals in championing a rapprochement between conservatism and modern Darwinism. More nuanced versions of this view can be found in the books of James Q. Wilson and in political theorist Larry Arnhart's closely argued book, *Darwinian Natural Right: The Biological Ethics of Human Nature*.[14] Both Wilson and Arnhart offer intriguing insights into the relationship between biology and morality, but it could be argued that they must redefine Darwinian biology in order to make room for traditional morality. Another approach is possible: Articulating a new theory of biology that can better account for the full range of human moral behavior. That is precisely what is being attempted by those advocating intelligent design.

More than a half-century ago, C. S. Lewis critiqued the consequences of scientific materialism for society in his penetrating manifesto *The Abolition of Man*. At the end of the book, Lewis called for a new kind of natural science that "[w]hen it explained it would not explain away. When it spoke of the parts it would remember the whole. While studying the *It* it would not lose what Martin Buber calls the *Thou*-situation."[15] Lewis wondered whether such a "regenerate science" was even possible, almost despairing that it was not. Those making the case for intelligent design today believe that such a science is indeed possible, and that they are helping to forge it. If they succeed, the benefits will spread far beyond the domain of science and will create a potent force for cultural renewal in the twenty-first century.

5

THE WORLD AS TEXT

Science, Letters, and the Recovery of Meaning

PATRICK HENRY REARDON

Patrick Henry Reardon, a senior editor at Touchstone *magazine, is pastor of All Saints Antiochian Orthodox Church in Chicago, Illinois.*

*P*hysics was one of my favorite pursuits in high school, where I spent many a happy afternoon leaning over some beaker to measure the specific weight of a portion of sandstone, or testing the mechanical advantage of some pulley or winch, or winding a magnetic field for a little motor I was constructing. Those were short-lived days, nonetheless, almost all of my later education being devoted to subjects usually subsumed under the heading of "letters": writing, history, literature and languages, theology, philosophy, and whatnot. Modern scientific scholarship has remained very much at the fringes of my mind.

Therefore, I was flattered when InterVarsity Press asked me to read the galleys of *Mere Creation*,[1] edited by William Dembski. Moreover, given the four decades or so that separate me from my last formal classes in biology, physics, or mathematics, I was further surprised at how much of the book I was able to understand, this latter circumstance doubtless to be ascribed to the extraordinary writing skill of its contributors rather than to any competence or intelligence of mine.

Professor Dembski and several of those same writers are found within the pages of this book. The importance of these chapters, whether taken severally or seen united by their common theme, will be obvious to all, nor is it thinkable that someone like me could add to their manifest authority. Rather, what I mean to do in the present article is to suggest, however faintly, some of the ways in which their shared thesis should recommend itself to the more general consideration of those not trained as scientists.

Science, after all, is not done in a vacuum. At every point, and most especially in its various starting points, scientific pursuit is based on and influenced by a host of popular, political, and other non-scientific presuppositions, cultural biases, and philosophical premises bequeathed from history, particularly the history of thought, religious thought most especially. To those who would separate science from its circumambient cultural world I recommend books like David C. Lindberg's *The Beginnings of Western Science*[2] and Lynn White's *Medieval Technology and Social Change*.[3] (Both of these are good reads, by the way.)

Indeed, the sheer specialization that has come to characterize the modern study of the physical sciences will tend to render the scientists themselves less aware of the philosophical, legal, historical, and even literary contexts in which their specific disciplines have evolved. After all, it is a rare university doctoral program nowadays that will permit a candidate to concentrate on chemistry and simultaneously take credit seminars in law or theater, or to major in medicine while minoring in music or Shakespeare. At the same time, theologians, historians, speculative philosophers, and men of letters generally, precisely because *non possunt omnia omnes,* tend to find themselves ever more isolated from the world of the pure sciences. This academic divorce is the subject of almost universal comment and widespread lamentation.

In what follows, therefore, I will suggest how the scientific thesis of these present writings may be related to the world of religion and letters.

Discerning a Text

Our earliest examples of that communicative device known as "writing" come from near the end of the fourth millennium before Christ and were bequeathed to us by a race that historians, taking their cue from the traditional name of the place, chose to call Sumerians. Thus we say that Sumerian was our first written language.

It is a most curious feature of these Sumerian texts that, though we are sure they represent a written language, we cannot decipher the first several centuries of them. Only the Sumerian texts from some five or so centuries later are we able to read. The reason for this is very simple. From this later period, about the middle to the end of the third millennium B.C., we happen to possess parallel texts written in a very ancient Semitic tongue, Akkadian, and we are able to use this latter language, much more firmly within the grasp of linguistic history, to make a reasonable reconstruction of later Sumerian vocabulary and grammar. Except for these Akkadian parallel texts, we would not be able to read Sumerian at all. There remain, nonetheless, those several earlier centuries of texts written in Sumerian that we are still unable to read.

With respect to these writings there is one thing, however, about which we are not in doubt—namely, that these are truly *writings*. That is to say, they really are symbols that carry an encoded meaning. They are, therefore, intentional; they are contrived signs crafted by human hands at the service of human minds in order to communicate with other human minds. They *say* something, whatever it may be. They form what is called "text," and it is a distinguishing feature of human beings that they can discern such a thing.

Whatever doubts, questions, or conjectures may be raised with regard to the meaning of these texts, it would be irresponsible to suppose that, because we are unable to decipher them, these graphic marks are without intentional meaning. To suppose that they are not texts, not writings, would fly in the face of the massive countervailing evidence.

And exactly what is this evidence? Just how do we know that certain configurations on a clay tablet, or a series of markings on a rock or stick, or even a string of beads forms a text? How are we able to distinguish these lines and configurations from random scribbling or doodling? What is it about them that causes us to presume, and even insist, that they are intended to convey specific information?

Likewise, why do we not take them to be simply artistic embellishments of some sort, similar to symmetric patterns that appear at far more distant periods of human history? Exactly what quality is it that prompts us to separate, as entirely new and special, the human markings that we find at Sumer and say, "Here is something truly different"?

Should we call that quality "intelligent design"? Well, perhaps, but I submit that this designation is really inadequate. After all, human beings produced very intelligent, highly crafted, graphic designs long before they began to write. If we were asked to explain exactly how a written text differs from these other designs, we might even say that writing shows, in fact, a certain lack of design, in the sense of not being perfectly regular and symmetric. Pieces of writing are marked by a sort of disparity, if you will, such as we would expect to find in a key, say, as distinct from a proportioned and uniform pattern.

To illustrate further: If I discover a sequence of graphic marks like /+0/+0/+0/+0 and so forth around the top of a vase, I am not likely to take this as an example of "text." Such a series of symbols is simply too regular to convey specific "information" of the sort I would expect in a spoken sentence. I readily suspect that, while such marks demonstrate an intelligent design, that design pertains to the realm of decorative art rather than of writing.

The latter contrivance, writing, though it certainly displays a structured form and consistency, is also characterized by a greater complexity, a subtle irregularity, and even a measure of unpredictability in its sequence. Its various symbols do not fall into place with the serial recurrence of a perfectly metric pattern, but in certain periodic conjunctions, even elaborate configurations, appearing at intervals not dictated by concerns of symmetry. Most people know when they are looking at a piece of writing; they are able to distinguish it from both random scribblings on the one hand and purely esthetic designs on the other.

However, writing does have this quality in common with artistic design—they are both recognized as "intentional," as distinct from random or haphazard. Both writings and artistic patterns are crafted with some purpose in mind, much as one would construct a tool. Indeed, this analogy with tools is worth considering. Ancient history has left us many examples of tools and implements whose distinct purpose we can only guess at. We look at certain ancient stone contrivances and can be perfectly sure that they were hand-held implements designed for a specific purpose, while others remain a mystery to us. That is to say, even though we are not sure of their purpose, we are nonetheless sure that they had a purpose, and this is how we recognize them as tools.

Such reflections demonstrate our ability to discern products of intelligence even when examining things that remain unintelligible to us. We human beings can detect the presence of "mind" and the intent to convey "information" even in those instances when we are unable to decipher exactly what that information is or what that mind is trying to say. Human reason has an innate capacity for recognizing the presence of rationality. We are able to know when we are, in fact, dealing with a *text*.

The Heavens Declare

An old tradition in Christian literature regards the created world itself as a kind of writing, a text for man's study. Indeed, because God created all things by his Word, it is not unknown for Christians to think of the created world as a sort of "first edition" of Holy Scripture. "The heavens declare the glory of God," wrote the psalmist, "day unto day uttereth speech, and night unto night sheweth knowledge....Their line is gone out through all the earth, and their words to the end of the world (Psalm 19:1, 2, 4)." The Apostle Paul would cite these very words, originally describing the message of creation, as illustrating the universality of the apostolic preaching (Romans 10:18). And did not the psalmist also compare the final dissolution of the heavens to the rolling up of a scroll?

Armed with this biblical warrant, earlier Christians turned to the world with a distinctly *exegetical* interest that Henri Cardinal de Lubac took some pains to document in the first volume of his *Medieval Exegesis*.[4] Examples of this interest are ubiquitous in Christian Latin literature. For instance, "these heavens," said St. Augustine, "these books, are the works of God's fingers," and St. Gregory the Great likened "the considered appearance of Creation" *(species considerata creaturae)* to "a sort of reading for our mind" *(quasi quaedam sit lectio menti nostrae)*. In the twelfth century, Alexander Neckam, the foster brother of Richard the Lion-Hearted, remarked that "the world is inscribed with the pen of God; for anyone who understands it, it is a work of literature." Similarly and slightly earlier, Rupert of Deutz had commented that, as the Word himself is "the architect of the world, he also ... composed ... Holy Scripture," and Herbert of Bosham even called the work of creation "a kind of corporeal and visible Gospel" *(velut quoddam evangelium corporale et visibile)*. Likewise, both Hugh and Richard of St. Victor wrote, "the whole of this sensible world is like a book written by the finger of God."

Perhaps no one spoke more often on this subject, however, than John Scotus Erigena in the ninth century. Commenting, for example, on the prologue of John's Gospel, he said that "the eternal light is proclaimed to the world in two ways, namely, Scripture and Creation." These form the double ladder of the divine ascent; commenting on the same Gospel, Erigena says later: "the first step in climbing the heights of virtue is the letter of Holy Scripture and the appearance of visible things, so that, once the letter has been read and creation examined, they may ascend, by the steps of correct reason, to the spirit of the letter and the rationality of creation."[5] Erigena speaks of creation and Holy Scripture as the two garments of Christ, a theme taken up three centuries later by Aelred of Rievaulx.

According to this traditional Christian view, the rationality of the universe, the clearly intentional disposition of its parts in relation to one another and to the whole, especially its relationship to man, is the sustaining subtext of the human narrative and the foundational context of poetry. Within the traditions of the Perennial Philosophy and the Christian faith, there is no need for man to formulate human meaning out of whole cloth. He is expected to presume, as more or less obvious, that the universe is possessed of a story line, that it is a rational and poetic place, in the sense of possessing a systematic coherence corresponding to the innate aspirations and essential structure of the human mind. The world and the intellect were made for one another.

Moreover, according to this traditional Christian and classical view, it is this textual quality of the created world that justifies our going to it in pursuit of literary types. For a writer like Dante, for example, if the structure of the world already contains an elaborate message, a message both poetical and rational, creation may itself be incorporated into man's literary endeavor as an active component. He finds in the world a pre-existing truth to be explored further by narrative. Beyond mere context, that is to say, creation provides also the structural types that give coherence to storytelling. Such was the persuasion of the age of faith. "The suggestion that truth, as well as the world, is out there," wrote Richard Rorty, "is a legacy of an age in which the world was seen as the creation of a being who had a language of his own." This conviction is part of the Christian birthright.

Literature and Random Evolution

Then came Charles Darwin. Roger Lundin, in his fine book *Emily Dickinson and the Art of Belief*,[6] describes the profound influence of Darwinian evolutionary theory on the writing of literature in the nineteenth century. Adopting the new persuasion that the physical world is a purely random setting, a heap of formless happenstance, devoid of purpose, and lacking in intentional design, some modern writers believed themselves faced with quite a new task—to create meaning out of whole cloth in a world where none existed. The shapeless world of nature could provide them with no "types," in the sense of permanent, preexisting intelligible patterns. On the contrary, nature, having nothing in itself to say, lay dormant and inert before their creative talents.

Such was the dilemma faced by Emily Dickinson, Lundin explains. Unable to think that the world is structured with an internal, intelligent coherence discernible to her thought, she felt obliged to undertake a literary task far more onerous than that facing earlier poets. Her vocation, she believed, was to "re-enchant" the world by her poetic imagination. When she observed the universe around her, she believed it was her literary calling to formulate certain literary *tropes* of her own, and these poetic associations would replace the *types* that earlier poets, such as Dante, believed to lie at the heart of the world. The difference between a trope and a type, Lundin explains, is that the latter is taken as an objective reality, whereas the former is the progeny of the poetic mind.

Surely there has grown a yawning skepticism in much of literature since the publication of *Origin of Species. Moby Dick,* written nearly a decade earlier, had already wrestled with the philosophical implications of a random universe. Darwin, however, provided the latter thesis with a scientific basis, postulating a "survival of the fittest" that would soon become Nietzsche's "will to power." Whether explicitly through Nietzsche or more indirectly through the cultural ambience created by the popular acceptance of evolutionary theory, Darwin's influence on modern literature appears to be immense. Sometimes, in fact, motifs drawn from an evolutionary perspective have formed the subject matter of literature itself; the novels of William Golding and the poetry of Ernesto Cardenal come to mind in this respect.

Darwin's darker shadow over the world of letters, nonetheless, seems more formal than material. No longer persuaded that they

76

inhabit an ordered, intelligible world, some modern storytellers from Henry James through James Joyce and Herman Hesse to Françoise Sagan have been reduced to examining mainly the current of their own subjectivity. Other authors, as diverse as Leo Tolstoy, Franz Kafka, Albert Camus, Samuel Beckett, Günter Grass, Alain Robbe-Grillet, and Kurt Vonnegut, have endeavored to redeem an indifferent and formless world by either somber humor, or cultivated oscitancy, or grotesque reveling, or even a desperate kind of compassion. Others yet, like John Steinbeck and Ken Kesey, adopting a dogmatic non-teleology as their point of departure, have struggled nonetheless to fill the resulting vacuum with biblical themes, to situate a secular-ized biblical narrative into a world not made by the biblical God. Still others, like Hemingway, have firmly asserted the virile self in the face of a menacing universe. Alas, the strength even of the manly Hemingway finally failed him, and his last violent act, though marked by a greater aggressive decisiveness, accomplished exactly what Vir-ginia Woolf did in her sad little walk out into the waves.

In addition, the formless and forlorn world of much of modern literature has been adequately matched, more recently by the new directions of postmodern literary criticism. At least in the popular mind, once science had decomposed the world, there was no rea-son why the world of letters should be exempt. Nor, as we all know, have the Sacred Letters themselves been spared the pervasive disso-lution wrought by nihilistic exegesis.

The Reassertion of Form

For all that, it is not the case that every major literary figure of these past two centuries has so completely swallowed, even reluc-tantly, the bitter pill of cosmic chaos, as Lundin shows Dickinson to have done. Gerard Manley Hopkins, for example, a poet arguably superior to Emily Dickinson, certainly had no such thing in mind when he proclaimed that the world is "charged with the glory of God," nor did the presumption of Darwin's random universe serve as cosmic backdrop to, say, *The Brothers Karamazov,* or *Les Misérables,* or *The Black Arrow,* or *Kristen Lavransdatter,* or *Perelandra.* From G. K. Chesterton and Wyndham Lewis through T. S. Eliot and Julien Greene to François Mauriac and Walker Percy, our own century has not lacked writers who refused to accept the evolutionary version of the cosmos. The universe of these authors differs hardly at all from the pre-Darwinian world of Alessandro Manzoni or Walter Scott.

And now science itself seems increasingly to justify their decision. The authors of *Mere Creation*, some of whom are found also in the pages of this book, demonstrate that the universe bears ample witness, bewildering in its complexity and wealth, to a coded but clear intentionality at the very core of things. The science of microbiology in particular demonstrates, within the subcellular components of all living entities, the ubiquitous presence of extremely intricate systems of information that a theory of random mutation simply cannot account for. That is to say, the discoveries of the pure sciences are forcing them to return to the ancient thesis that the structure of the world forms a kind of *writing*. The current study of chemistry, biology, and astrophysics testifies that the information within the universe presents us with a *text*.

Some period of time, doubtless, will be required for this scientific rediscovery to permeate the popular mind. Ironically, it may take even more time for it to permeate the academic mind, a later disposition to mental lethargy being one of the demonstrable fruits of academic tenure.

Indeed, one suspects that modern philosophers and theologians, who have created the most elaborate and improbable theories to reconcile their religious ideas with an outdated state of science, may be the last segment of society to get on board, so long have they been wandering around in the fog of Bergson's *élan vital*, striving to attain Whitehead's *actual occasions*, and hoping to reach Teilhard's *omega point*. The depth of this philosophical resistance to the recent testimony of science is evidenced, for example, in Robert Pennock's *Tower of Babel: The Evidence Against the New Creationism*, which, in spite of its promising subtitle, introduces not a shred of new evidence in support of Darwinism, unless under that heading we include his puerile remarks about the variety of sexual postures employed by other primates as confirmation of their biohistorical relationship to human beings. Largely ignoring the strong arguments advanced against Darwinism from the direction of biochemistry and microbiology, Pennock spent most of his efforts on the purely rhetorical and deceptive task of painting intelligent design theorists as simply a new species of creation scientists.

Meanwhile, it is my personal hope that these positions newly adopted by scholars in the scientific community, when they do reach the larger world, will lead in turn to a renewal of philosophy and humane letters, and that an enhanced confidence in the ordered structure of physical reality will afford men and women a more assured, firmer stride in the paths of narrative and poetic composition. Actually, I have no doubt that this will be the case, at least after my time, and I cherish the suspicion that future students of literary history, not so terribly far down the road, may look back to these past two centuries as a somewhat weird period, during which an extraordinary multitude of singularly disturbed authors composed an inordinate number of very bizarre and disquieting books. "Yes," their teachers will be obliged to inform them, "a lot of people back in those unfortunate days had gotten it into their silly heads that the whole world and everything in it had somehow evolved by accident, you see. It was all rather strange."

6

GETTING GOD A PASS

Science, Theology, and the Consideration of Intelligent Design

JOHN MARK REYNOLDS

John Mark Reynolds, Ph.D. (philosophy, University of Rochester), is associate professor of philosophy at Biola University, where he founded and directs the Torrey Honors Institute (a Great Books program for under-graduates). He recently coedited with J. P. Moreland Three Views on Creation and Evolution[1] *and is currently completing a book about the history of ancient philosophy. He is a member of the Orthodox Church (Western Rite).*

*M*y son Ian loves Disneyland. Recently he loved it so much that he decided to stay. Our four-year-old wandered away from us in a store and we spent the next twenty minutes looking for him. When he was discovered, appropriately enough, near *Mr. Toad's Wild Ride*, we were relieved. We found him sitting in the middle of a group of doting Disneyland cast members. He was wearing new mouse ears with his name on them and was carrying a brand-new Mickey Mouse plush toy. His brother and sisters were not amused.

We were so happy to see him that we could not punish him. Carrying him out to the car, his mother began to quiz him on what had taken place in the twenty minutes he had been out from under her

80

care. Thriving on the attention, Mr. Ian responded by telling an expansive tale of his adventures with the big men from security who had tried to catch him.

Finally, her gratitude overflowing, Hope said, "Well, Ian, we are so thankful to God for bringing you back to us. Mommy and Daddy prayed and Jesus heard us."

Ian looked at his mother with the slightly bemused look usually reserved for the mentally disabled. "Mom," he said with care, "Stop it. You are ruining the story."

Hope was shocked. "What do you mean, Ian?"

Mr. Ian replied, "God had nothing to do with the story. The story is about me."

"But Ian," she replied in vain, "God is everywhere."

Our four-year-old rolled his eyes and explained, "Mom, to get into Disneyland you must have a pass. God does not have a pass to Disneyland."

Those Christians involved in the boycott of Disney might be tempted to agree with my son's statement. In the end, however, all Christians have to agree that while God may not have a pass to Disneyland, he is certainly there. It is understandable when a small child makes such a mistake. After all, superficially God had nothing to do with his "rescue." Because of his simple view of causation, God was not necessary to explain the events in the park that day.

The Theistic Naturalist

What is excusable in a child is, however, inexcusable when it comes to adults, especially ones who should know better. Many Christians who know, in the abstract, that God is everywhere and can do anything become shifty and obscure when it comes to God's detectable involvement in the natural world. These Christian scientists, theologians, and philosophers think that the addition of theism to science ruins the story.

Why is this so? The great human enterprise of science does not seem to need God in order to generate its research programs. Traditional scientists never have a need for the divine or supernatural hypothesis. In the West, some Christians have made seemingly false predictions about the nature of the world based on their understanding of divine revelation. Many Christian academics, therefore, have decided that any philosophy of science or theory of science that puts "God in the dock" is a bad move.

Of course, those Christian thinkers who see things this way do not embrace the God-denying philosophy of naturalism. Instead, they embrace a "methodological naturalism." This technique allows the scientist to believe that God exists, but to do his work as a scientist as if God did not exist. God may have done things in space and time, but (fortunately for the career path of these Christians) the Almighty never did anything that could be confirmed by unbelievers in a laboratory. The theist and the nontheist can do their work in the same manner with the same practical expectations.

The result is a theism that fits into the epistemological gaps left to it by the scientists. Only where natural science does not (and seemingly cannot) speak is theology free to make pronouncements. The difficulty is that the "off limits" areas keep shrinking. Naturalists of the more robust variety are not so bashful in pressing the claims of their philosophy. What does a theist who is also a methodological naturalist do? He retreats. New meanings are discovered for old doctrines. People who will not go along are called "fundamentalists." Peace is kept with the secular scientific guild. Not surprisingly, these theistic naturalists find they have less and less to say theologically that is, even in principle, verifiable by empirical means. In a religion that began with the bold proclamation of an empty tomb and a risen Lord vouched for by hundreds of witnesses, this is an odd position.

The God of the Gaps

The retreat of the theistic naturalist's God into ever-shrinking gaps has been going on for the last century. An excellent recent example is the theistic naturalists' reaction to the continued growth of fully naturalistic psychology. Traditional Christians have almost universally proclaimed their belief in an immortal soul, distinct from the brain. People have souls. If people have souls not made of matter and energy, an important limitation is placed on a naturalistic science.

The difficulty for the theistic naturalist is that naturalistic science was not content to leave psychology to theologians. Naturalistic methodology, which the theistic naturalist has already accepted, has been applied to human beings. Academic fashion has firmly decreed that there is no place for the soul in their psychological theorizing.

What were the theistic naturalists to do? Suddenly they discovered that the idea of a nonmaterial soul is theologically controversial. The blame for the old, discarded notion of a soul may be placed on Descartes or it may be placed on Plato. It does not matter. The new trend in theistic naturalism is to rid people of souls and leave them with brains. Mind, for these thinkers, can be explained as the product of matter and energy. There is no "ghost" in the machine. People think with their brains. The theistic naturalist then argues that this is how biblical revelation and Christian theology should have been understood all along.

Of course, this leaves God, who has no brain, as a great exception for the theistic naturalist. In his retreat from dualism, the materialistic theist will often ask how a nonmaterial soul could influence a material body. This putative problem, when combined with the findings of naturalistic psychology, seems sufficient to eliminate the soul. It is up to theology to reinterpret itself in the light of this "new" idea.

There is a problem with this happy accommodation. Christian theology insists that a nonmaterial God has interacted with a material cosmos. The theistic naturalists have choked on a metaphysical gnat, while swallowing the camel.

Theistic naturalists are not too concerned about consistency on this issue, however. Science, thank God, will never be able to give a naturalistic account of God and his actions. He simply would not stand for the examination. God is left alone, very much alone. He becomes the subject of a personal piety that is beyond evidence.

To their credit, many of these academics wish to remain traditional Christians. They wish to avoid deism. They have seemingly pushed God very far out of the picture. What do they do to avoid an inactive God? First, they postulate God's direct action in salvation history. Most theistic naturalists would accept the literal, physical resurrection of Jesus. This is, of course, commendable.

There is a limit to this traditionalism, however. Many of the Old Testament portions of salvation history do not fit contemporary natural science. Not surprisingly, these parts are often reinterpreted as non-historical. The happy result for the theistic naturalist is that exactly those stories most open to modern verification are declared purely theological and off limits. And stories, like the story of the resurrection, that are no longer likely to be challenged on scientific grounds are allowed historical content. Put another way, God uses "primary causation" when no scientists are looking. He seems to use "secondary causation" the rest of the time. This may be a logically possible move, but it is not an inspiring one.

God continuously sustains the universe. No orthodox Christian can deny this theological fact. By agreeing with this cardinal doctrine, the theistic naturalist hopes to protect his views from deism. The problem is trying to find out what the theistic naturalist means by "sustains." Traditionally, Christians have understood this to mean that God directly causes the cosmos to exist at any given moment. It is a property of matter and energy that it cannot exist outside the power of God.

If this doctrine is true, then methodological naturalism must be abandoned in physics. Physics certainly attempts to answer questions about the existence of matter and energy. If God is necessary to answer even one of these questions, then methodological naturalism will blind science to the truth. Logically, this would mean that either methodological naturalism or theological commitment must go. In trying to avoid deism, the theistic naturalist destroys his or her own position. Generally, however, the theistic naturalist is content to leave God's "sustaining" as a vague, religious platitude. A good question to ask the theistic naturalist is this: "What does God do that can be verified, at this time, in the natural world?"

Signposts to God

Even the pagan can see the hand of God when he or she looks at the natural world. Every Christian tradition is bound to such a view. The Gospels did not simply celebrate a subjective Easter event; they pointed to the space-time events of the Passion Week. This history is affirmed at every liturgy within the ancient tradition of the church.

The Christian need only look at Romans 1:18–23 to see that this tradition is biblical:

> For the wrath of God is revealed from heaven against all ungodliness and unrighteousness of men, who hold the truth in unrighteousness; Because that which may be known of God is manifest in them; for God hath shown it unto them. For the invisible things of him from the creation of the world are clearly seen, being understood by the things that are made, even his eternal power and Godhead; so that they are without excuse: Because that, when they knew God, they glorified him not as God, neither were thankful; but became vain in their imaginations, and their foolish heart was darkened. Professing themselves to be wise, they became fools, and changed the

glory of the incorruptible God into an image made like to corruptible man, and to birds, and four-footed beasts, and creeping things.

This passage has traditionally been understood to mean that even non-Christians could see the reality of God in the cosmos. This reality is not hidden behind a veil of materialistic science; it is plain to all. St. John Chrysostom says in his commentary on Romans:

But do thou make it good, and show me that the knowledge of God was plain to them, and that they willingly turned aside. Whence was it plain then? Did He send them a voice from above? By no means. But what was able to draw them to Him more than a voice, that He did, by putting before them the Creation, so that both wise, and unlearned, and Scythian, and barbarian, having through sight learned the beauty of the things which were seen, might mount up to God.[2]

For the traditional Christian, nature acts as a signpost to God. For the theistic naturalist, the signpost is only readable to the enlightened few whose theology allows them to strip the materialism from their science.

Indeed, the ancients show the truth of Chrysostom's words. Plato, Aristotle, and most of the other pagan philosophers looked for Mind or some nonmaterial cause in the cosmos. Plato himself in *Laws* X rejects the idea that the cosmos could come into being from purely naturalistic causes. With the possible exception of certain philosophers like Lucretius, Christianity had no problem arguing for the existence of a rational Creator. In fact, this situation was true for most of human history. Before the nineteenth century, the village atheist was considered a crank. The argument for design in the cosmos was simply too potent to be ignored. It is forgotten that even as robust a religious group as the English Puritans viewed the advance of science as leading to an increase in the Christian religion. As Richard Dawkins points out in his *Blind Watchmaker,* only Darwin made it possible to be an intellectually fulfilled atheist.

God can create the world in any manner he chooses. Theistic naturalists often assert that evolution and methodological naturalism can be made compatible with Christianity. This is true. Almost any two ideas can be made compatible with sufficient motivation. The issue is much simpler in reality. *Is Darwinism true?* Theistic naturalists never want to discuss this issue.

Perhaps a person might object that complex theories like Darwinism could not be described as true or false. Theories always have problems after all. A more "sophisticated" question might be *"Given the evidence now in hand, is Darwinism plausible or implausible?"* That is fine, as far as it goes. If the "evidence now in hand" includes theological evidence, then traditional Christians find the question appropriate. Too often, however, theistic naturalists do not want to count the propositions of theology as "knowledge" in this sense. They wish to relegate what is known about the divine to a different sphere. As J. P. Moreland has demonstrated in *Christianity and the Nature of Science*,[3] such lines of demarcation cannot be drawn.

Isn't this all just a concern of those nasty fundamentalists? I know Christians from fundamentalist backgrounds who have joined one of the historic churches and are relieved to think that the church has nothing to say about such issues. Such converts feel that, since they no longer have to buy into a wooden Genesis literalism, problems about Darwinism can be left to the side.

It is true that the church has no dogma on this issue. This is an area where theological and scientific speculations have been and should remain unfettered. On the other hand, such freedom does not mean that just anything will do. Some ideas are implausible in light of Christian tradition, biblical revelation, and reason. Darwinism did not develop as a result of careful Christian thought. The church has not prospered where Darwinism has flourished. This should be motivation enough to ask if a rational person is compelled to believe it. If the evidence for it were overwhelming, then the rational Christian would be forced to modify his views. The evidence for Darwinism is not overwhelming. Phillip Johnson has carefully demonstrated this in his book *Darwin on Trial*.[4] The orthodox Christian has motive and opportunity to free himself from the shackles of this Victorian creation myth.

The issue, when carefully understood, is not whether Genesis should be interpreted literally. It is whether science or theology will be considered the paramount source of knowledge in the culture. In the West, from Galileo to the present, religious people have tried to argue that nature and revelation are coequal domains of knowledge. They have tried such lines as, "The Bible tells us how the heavens came to be. Science tells us how they go."

The result has been a disaster. How can a reasonable person know that the Bible or the church is correct? Science, which includes the realm of historical evidence, has been removed as a possible check on the truth of Scripture or church tradition. Functionally, theology is relegated to the world of subjective experience. It cannot be verified by the checks of common external experience. The church is left in the same position as the Mormon who appeals to the "burning in his bosom" to justify his faith. But those who accept the Virgin's fruitful womb and the empty tomb cannot stand such a separation. We believe that such things—incarnation and resurrection—happened in space and time. This is not just theological speculation, but something we know.

The Incarnation and Theistic Evolution

On occasion I am privileged to serve at the altar of our parish. The entire service is a reminder of the incarnation. As the priest lifts the bread and wine to bless it, I see the supreme mystery of the church. Then I taste of that feast. On every side of me are the images that the church has fought to preserve. These images provide a way for my eyes to see heaven. The smell of incense fills the air. The sound of bells draws my attention back to the altar as I kneel there. My body and soul are united in the actions of the celebration. At the end of the service we kneel when the priest repeats John 1:14, "And the Word was made flesh, and dwelt among us, and we beheld his glory, the glory as of the only begotten of the Father, full of grace and truth."

From the first to the last, in the historic liturgy of the church the mystery of the Passion is connected to the mystery of the Word that became flesh. Christians have never hated matter and energy, for our God took on human form. The services of the historic church are rich with the sights, smells, and tastes of the world, because the world our God created is good.

Orthodox Christianity recognizes the importance of the soul and each of the senses to worship. Each is ministered to and brought into union in the Eucharistic feast. The church does not deny that truth is propositional, but also does not forget that Truth himself is a man. The liturgy brings into union heaven and earth. It is the church's greatest gift to civilization apart from the gospel itself.

This theology of the incarnation helped make modern science possible. The pagan Greeks had insufficient motivation for continuous study of the material world. They veered between thinking nature divine and thinking it wicked. Some pagans elevated the natural world to divine status. One does not do experiments on God.

The more sophisticated Greek thinkers saw the dual nature of reality. Too often, though, they slipped into the view that matter was bad or beneath them. Aristotle and a few other philosophers tried to challenge this, but with little success. Having correctly identified the two realms, heaven and the cosmos, the Greeks were unable to keep from despising the cosmos. Science was impossible in such a situation.

When the Word became flesh, it became possible to see the creation as good without thinking of it as God. Matter was made worthy of serious study. It had provided flesh for God. Just as every mother can find transcendent validation of her calling in the actions of the mother of God, so the scientist finds purpose to his calling in the incarnation.

The liturgy allows even the pagan to see the beauty of God and his Word. The icon allows even the non-Christian a window to heaven. The story is told that Russia was converted after the combination of physical and spiritual beauty overwhelmed their ambassadors in the Great Church of the Holy Wisdom in Constantinople. Even the pagan, as St. John points out, can see the Word when he puts on human flesh. Our reaction to his light may be to try to snuff it out or worship his glory, but his glory can be seen.

Theistic naturalism destroys this image. It makes nature a closed book. While still maintaining that it was created by God, the theistic naturalist holds that a pagan can look at it without seeing his glory. The person of science can handle creation without seeing that it is a creation. The theistic naturalist would have the Christian adopt rules for science that would prevent the discernment of glory. It is as if a liturgical church were to be stripped bare of all physical objects. In such a church, the liturgy would be disconnected from the physical world and become a matter only for an interior mental world. This would be a disaster.

Intelligent Design Lets God In

Such a separation is an equal disaster in science and it is unnecessary. Much has changed since the days of Darwin and the design arguments of William Paley. Darwinism, however, has not managed to keep up with the explosion of data that is now being brought to our attention. It remains the same shopworn nineteenth-century relic, with cobbled-on modifications, that it has always been. Why should the church bother to try to save the creation myth of naturalists?

A revolution is taking place all around us. Scientists like Michael Behe, author of *Darwin's Black Box*, are showing that there is massive evidence of design at the biochemical level. William Dembski, in what may prove to be one of the most important philosophic clarifications of the late twentieth century, has refined the very notion of design. Intelligent design is rapidly becoming an important intellectual movement.

It is not hard for the believer to move from the idea of design in creation to the God of Christian faith. This work has been done, and is being done, by philosophers like J. P. Moreland and William Lane Craig. Increasingly it is becoming obvious that the person who looks at nature can see direct evidence for creation. Nature itself, even in its fallen state, remains an icon of the glory of God.

Intelligent design, though not a theological cause, is a movement more compatible with Christianity than any form of naturalism, be it metaphysical, methodological, or theistic. Intelligent design allows the church to believe what she has always believed, that even the pagan can see the hand of God if he or she looks at nature. Intelligent design adopts an open philosophy of science that is not afraid to see the work of an agent in the natural world. Theistic naturalism is a crabbed and limited affair that allows God to do only what secular science will allow him to do. The theistic naturalists should be allowed the freedom to continue their work, but they should not be surprised if neither the church nor the secular culture is very interested.

My son Ian did not believe God had a pass to Disneyland. He did not look for God's action, even when others could see that God had acted. Ian had adopted too small a worldview. He needs to realize that God is everywhere, even in Disneyland. If Christians will closely examine the findings of the intelligent design movement, they will see that God never needed a pass to enter the world of science. His very image has been there all the time.

7

DARWIN'S BREAKDOWN

Irreducible Complexity and Design at the Foundation of Life

MICHAEL J. BEHE

Michael Behe, Ph.D. (biochemistry, University of Pennsylvania), is professor in the department of biological sciences at Lehigh University in Pennsylvania and the author of Darwin's Black Box: The Biochemical Challenge to Evolution. *His biochemical research has been funded by the National Institutes of Health and the National Science Foundation. He is a member of the Biophysical Society and the American Society for Molecular Biology and Biochemistry. This article is adapted from essays published in* Ethics and Medics *and* Creation and Evolution: The Proceedings of the October 1997 ITEST Workshop.

*T*he topic of evolution is both fascinating and vexing. Fascinating, because it involves questions of who we are, how we got here, and how we relate to the world around us. Vexing, because it provokes arguments and political battles. The battles are all the more frustrating when fought by polarized factions with preset agendas such as, on the one hand, defending a literal interpretation of the Bible, or, on the other, keeping any hint of transcendence out of the public schools.

Many people, discouraged by the bickering, are tempted to ignore the topic. For a Christian, this is a mistake. In his letter to the Pontifical Academy of Sciences several years ago (*L'Osservatore Romano*, October 30, 1996), Pope John Paul II wrote that evolution is "an essential subject which deeply interests the church, since revelation, for its part, contains teaching concerning the nature and origins of man." An informed Christian, engaged with the world, should have a basic understanding of the facets of evolution that are of particular importance to the faith.

In his letter the pope made several statements that seemed on the surface to conflict. He acknowledged that evolution is "more than a hypothesis" and that "this theory has been progressively accepted by researchers, following a series of discoveries in various fields of knowledge." He also noted, however, that "rather than the theory of evolution, we should speak of several theories of evolution." In part the several theories have to do with "the different explanations advanced for the mechanism of evolution." How can evolution be so well-supported that it is more than a hypothesis and yet there still be uncertainty about its mechanism?

The difficulty arises because the word *evolution* can be used in different senses, and equivocation can easily confuse people. In one sense evolution just means common descent—that living creatures are all related to a common ancestor. It is in this sense, I think, that the pope meant that evolution is well-supported. Common descent is a possible explanation for the *similarities* among species, and modern science has shown many similarities, especially at the molecular level, that were unknown to earlier scientists. Explaining similarities, however, is the easy part. You just have to say that some features remained the same. Explaining the many profound *differences* between organisms is the hard part.

In another sense evolution is sometimes used to mean Darwin's particular theory. Darwin proposed the theory of natural selection to account for the differences between organisms. Darwin saw that there was variation in all species, and he reasoned that animals whose random variation gave them an edge in the struggle to survive would tend to leave more offspring than others. If the variations were inherited, then the characteristics of the species would change over time. In other words, Darwin proposed a mechanism to drive evolution. Although it clearly can explain relatively small changes, the sufficiency of Darwin's mechanism to account for larger, more complex changes in organisms remains in question. Pope John Paul II apparently had in mind the doubts about Darwin's mechanism when he referred to "several theories of evolution."

The Professor and the Cardinal

Why do such arcana matter? To many Christians, the problem with Darwin's theory is in the single word *random*. Ever since the theory was first proposed, persons antagonistic to the church, including some prominent scientists, have aggressively asserted that the randomness is not merely epistemic; it is ontological. In other words, they claim that science sees no purpose in living things because there is no purpose, and therefore there is no God. For example, the Oxford zoologist Richard Dawkins has remarked that "The universe we observe has precisely the properties we should expect if there is at bottom no design, no purpose, no evil and no good, nothing but pointless indifference."[1] Clearly such assertions go well beyond the domain of science. Nonetheless, because science has considerable authority in our culture, scientists are accorded great respect even when they are dispensing bad philosophy.

Although in his letter the pope did not address the randomness assumed by Darwin's mechanism, his close advisor, Cardinal Joseph Ratzinger, did discuss it in a little book entitled *In the Beginning: A Catholic Understanding of the Story of Creation and the Fall*.[2] There he wrote:

> Let us go directly to the question of evolution and its mechanisms. Microbiology and biochemistry have brought revolutionary insights here. . . . [W]e must have the audacity to say that the great projects of the living creation are not the products of chance and error. . . . [They] point to a creating Reason and show us a creating Intelligence, and they do so more luminously and radiantly today than ever before.[3]

Let us note three things about the cardinal's argument. First, he is claiming that life is the result of intelligent purpose, not ontologically random events. Second, he bases this claim on physical evidence (i.e., the great projects of the living creation that point to a creating Reason), not on scriptural or theological arguments. Finally, he implies that biochemistry, the study of the molecular basis of life, provides particularly strong support for this view. He has a formidable point.

Darwin Said It Best

In 1996 I wrote *Darwin's Black Box: The Biochemical Challenge to Evolution,* whose main point Cardinal Ratzinger anticipated ten years earlier. *Black box* is a term used in science for a machine or system that does interesting things, but whose inner workings are unknown. To Darwin and other nineteenth-century scientists, the cell was a black box. Darwin advanced his theory in an age when the fundamental mechanisms of life were completely obscure.

Modern science has found that, far from the simple "protoplasm" that many nineteenth-century scientists believed, the cell contains ultra-sophisticated molecular machines. The assumption that the basis of life is simple has turned out to be the polar opposite of the case. Now that modern science has unveiled the surprising complexity of molecular life, how can we decide if Darwin's theory can account for it? It turns out that Darwin himself set the standard. In his *Origin of Species*, he acknowledged that: "If it could be demonstrated that any complex organ existed which could not possibly have been formed by numerous, successive, slight modifications, my theory would absolutely break down."[4]

But what type of biological system could not be formed by "numerous, successive, slight modifications"? A system that is *irreducibly complex*. Irreducible complexity is just a fancy phrase I use to mean a single system that is composed of several interacting parts, where the removal of any one of the parts causes the system to cease functioning.

Let's consider an everyday example of irreducible complexity: the humble mousetrap. The mousetraps that my family uses consist of a number of parts (Figure 1). There are: (1) a flat wooden platform to

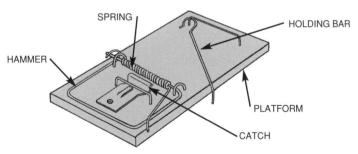

Figure 1. A household mousetrap. The working parts of the trap are labeled. If any of the parts is missing, the trap does not function.

act as a base; (2) a metal hammer, which does the actual job of crushing the little mouse; (3) a spring with extended ends to press against the platform and the hammer when the trap is charged; (4) a sensitive catch that releases when slight pressure is applied; and (5) a metal bar that connects to the catch and holds the hammer back when the trap is charged. Now you can't catch a mouse with just a platform, add a spring and catch a few more mice, add a holding bar and catch a few more. All the pieces of the mousetrap have to be in place before you catch any mice. Therefore the mousetrap is irreducibly complex.

An irreducibly complex system cannot be produced directly by numerous, successive, slight modifications of a precursor system, because any precursor to an irreducibly complex system that is missing a part is by definition nonfunctional. An irreducibly complex biological system, if there is such a thing, would be a powerful challenge to Darwinian evolution. Since natural selection can only choose systems that are already working, then, if a biological system cannot be produced gradually, it would have to arise as an integrated unit for natural selection to have anything to act on.

Let me add a word of caution. Demonstration that a system is irreducibly complex is *not* a proof that there is absolutely no gradual route to its production. Although an irreducibly complex system can't be produced directly, one can't definitively rule out the possibility of an indirect, circuitous route. However, as the complexity of an interacting system increases, the likelihood of such an indirect route drops precipitously. And as the number of unexplained, irreducibly complex biological systems increases, our confidence that Darwin's criterion of failure has been met skyrockets toward the maximum that science allows.

The Cilium

Mousetraps are one thing, biochemical systems are another. So we must ask, are any biochemical systems irreducibly complex? Yes, it turns out that many are. A good example is the cilium. Cilia are hairlike structures on the surfaces of many animal and lower plant cells that can move fluid over the cell's surface or "row" single cells through a fluid. In human beings, for example, cells lining the respiratory tract each have about two hundred cilia that beat in synchrony to sweep mucus towards the throat for elimination.

94

What is the structure of a cilium? A cilium consists of a bundle of fibers called an axoneme. An axoneme contains a ring of nine double "microtubules" surrounding two central single microtubules. Each outer doublet consists of a ring of thirteen filaments fused to an assembly of ten filaments. The filaments of the microtubules are composed of two proteins called alpha and beta tubulin. The eleven microtubules forming an axoneme are held together by three types of connectors: outer doublets are joined to the central microtubules by radial spokes; adjacent outer doublets are joined to each other by linkers of a highly elastic protein called nexin; and the central microtubules are joined by a connecting bridge. Finally, every doublet bears two arms, an inner arm and an outer arm, both containing a protein called dynein.

Although even this seems complex, a brief description can't do justice to the full complexity of the cilium, which has been shown by biochemical analysis to contain about two hundred separate kinds of protein parts.

But how does a cilium work? Experiments have shown that ciliary motion results from the chemically powered "walking" of the dynein arms on one microtubule up a second microtubule so that the two microtubules slide past each other (Figure 2). The protein

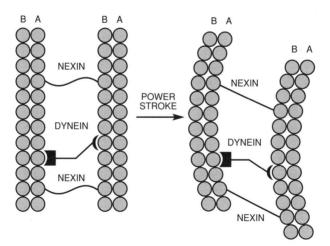

Figure 2. Schematic drawing of part of a cilium. The power stroke of the motor protein, attached to one microtubule, against a neighboring microtuble causes the fibers to slide past each other. The flexible linker protein, nexin, converts the sliding motion to a bending motion.

crosslinks between microtubules in a cilium prevent neighboring microtubules from sliding past each other by more than a short distance. These crosslinks, therefore, convert the dynein-powered sliding motion to a bending motion of the entire axoneme.

Now, let us consider what this implies. What components are needed for a cilium to work? Ciliary motion certainly requires microtubules; otherwise, there would be no strands to slide. Additionally we require a motor, or else the microtubules of the cilium would lie stiff and motionless. Furthermore, we require linkers to tug on neighboring strands, converting the sliding motion into a bending motion, and preventing the structure from falling apart. All of these parts are required to perform one function: ciliary motion. Just as a mousetrap does not work unless all of its constituent parts are present, ciliary motion simply does not exist in the absence of microtubules, connectors, and motors. Therefore, we can conclude that the cilium is irreducibly complex—an enormous monkey wrench thrown into its presumed gradual, Darwinian evolution.

A Complex Delivery System

Another example of irreducible complexity is the system that targets proteins for delivery to subcellular compartments. The eukaryotic cell contains a number of subcellular compartments for specialized tasks, like rooms in a house. These include lysosomes for digestion, Golgi vesicles for export, and others. Unfortunately, the machinery for making proteins is outside these compartments, so how do the proteins that perform tasks in subcellular compartments find their way to their destination? It turns out that proteins that will wind up in subcellular compartments contain a special amino acid sequence near the beginning called a "signal sequence." As the proteins are being synthesized, a complex molecular assemblage called the signal recognition particle (SRP) binds to the signal sequence. This causes synthesis of the protein to halt temporarily. During the pause in protein synthesis the SRP binds the trans-membrane SRP receptor, which causes protein synthesis to resume and which allows passage of the protein into the interior of the endoplasmic reticulum (ER). As the protein passes into the ER the signal sequence is cut off.

For many proteins the ER is just a way station on their travels to their final destinations (Figure 3). Proteins that will end up in a lyso-

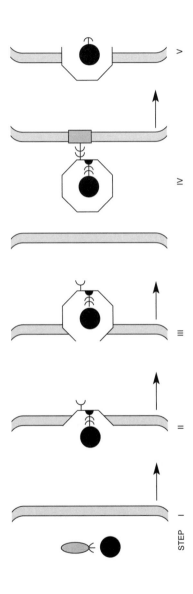

Figure 3. Transport of a protein from the ER to the lysosome. Step I: A specific enzyme (gray oval) places a marker on the protein (black sphere). This takes place within the ER, which is delimited by a barrier membrane (bar with ends curving to the left). Step II: The marker is specifically recognized by a receptor protein and the clathrin vesicle (hexagonal shape) begins to form. Step III: The clathrin vesicle is completed and buds off from the ER membrane. Step IV: The clathrin vesicle crosses the cytoplasm and attaches through another specific marker to a receptor protein (dark gray box) on the lysosomal membrane (bar with ends curving to the right). Step V: Through a series of several more steps, the clathrin vesicle fuses with the lysosomal membrane and releases its cargo.

some are enzymatically "tagged" with a carbohydrate residue called mannose-6-phosphate while still in the ER. An area of the ER membrane then begins to concentrate several proteins; one protein, clathrin, forms a sort of geodesic dome called a coated vesicle, which buds off from the ER. In the dome there is also a receptor protein that binds to both the clathrin and to the mannose-6-phosphate group of the protein that is being transported. The coated vesicle then leaves the ER, travels through the cytoplasm, and binds to the lysosome through another specific receptor protein. Finally, in a maneuver involving several more proteins, the vesicle fuses with the lysosome and the protein is at its destination.

During its travels our protein interacted with dozens of macromolecules to achieve one purpose: its arrival in the lysosome. Virtually all components of the transport system are necessary for the system to operate, and therefore the system is irreducible. The consequences of a gap in the transport chain can be seen in the hereditary defect known as I-cell disease. It results from a deficiency of the enzyme that places the mannose-6-phosphate on proteins to be targeted to the lysosomes. I-cell disease is characterized by progressive retardation, skeletal deformities, and early death.

Detection of Design

Many other examples of irreducibly complex cellular systems could be cited. In the face of such unexpectedly sophisticated machinery, what have scientists said? A number of prominent scientists have come to doubt that natural selection can account for them. The shortcomings of Darwinian explanations have been noted by Stuart Kauffman, Lynn Margulis, Brian Goodwin, James Shapiro, and others. Each of them has proposed an alternative to Darwinism that appeals to some sort of blind, general law. To date, however, none of the alternatives has had much success.

While I agree with these scientists that many biochemical systems can't be explained by natural selection, I differ in the alternative I offer. I argue that the systems show strong evidence of design—purposeful, intentional design by an intelligent agent.

How do we detect design? The criteria for deciding that something has been designed have recently been outlined with mathematical and philosophical rigor by William Dembski in *The Design*

Inference. Here I would like to give a simple, intuitive criterion for suspecting design in discrete physical systems. In these cases design is most easily apprehended when a number of separate, interacting components are ordered in such a way as to accomplish a function beyond the individual components. To illustrate, think of the Far Side cartoon by Gary Larson in which an exploring team is going through a jungle. The lead explorer is pulled up and left dangling from a tree by a vine wrapped around his foot, and has been skewered by wooden spikes aimed precisely at the position in the air where the vines pulled him. A companion turns to another and confides, "That's why I never walk in front."

Now, every person who sees the cartoon knows immediately that the trap was designed. But how does one know that? How does the audience apprehend that the trap was designed? One can tell that the trap was designed because of the way the parts interact with great specificity to perform a function. Like the mousetrap in Figure 1, no one would mistake the cartoon trap for an accidental arrangement of parts.

I argue that many biochemical systems were designed by an intelligent agent. Our apprehension of the design of the cilium or intracellular transport rests on the same principles as our apprehension of the design of the jungle trap: the ordering of separate components to achieve an identifiable function that depends sharply on the components.

A Deliberate Resistance

In all the reviews of my book that I am aware of, no one claims that the biochemical systems I describe have already been explained by science. James Shreeve, reviewing the book for the *New York Times*[5], says, "Mr. Behe may be right that given our current state of knowledge, good old Darwinian evolution cannot explain the origin of blood clotting or cellular transport." In the *National Review*[6] microbiologist James Shapiro of the University of Chicago writes, "There are no detailed Darwinian accounts for the evolution of any fundamental biochemical or cellular system, only a variety of wishful speculations."

In *Nature*[7] University of Chicago evolutionary biologist Jerry Coyne, although very unfriendly to the concept of intelligent design, states, "There is no doubt that the pathways described by Behe are dauntingly complex, and their evolution will be hard to unravel. . . . We may forever be unable to envisage the first proto-pathways." In *New Scientist*[8] Andrew Pomiankowski writes, "Pick up any bio-chemistry textbook, and you will find perhaps two or three refer-ences to evolution. Turn to one of these and you will be lucky to find anything better than 'evolution selects the fittest molecules for their biological function.'" So apparently everyone at least agrees that complex biochemical systems have yet to be explained.

However, none of the reviewers who are biologists agrees with the conclusion of intelligent design. Shreeve worries, "Shouldn't we leave something for our children and grandchildren to puzzle out besides which systems in the cell are intelligently designed and which are not? Because something is beyond our understanding today does not mean it will be beyond theirs." Shapiro laments, "Sadly, despite its valuable critique of an all-too-often unchallenged orthodoxy, *Darwin's Black Box* fails to capture the true excitement of contem-porary biology because it is fighting the battles of the past rather than seeing the vision of the future." Coyne sniffs, "It is not valid, however, to assume that, because one man cannot imagine such path-ways, they could not have existed." And Pomiankowski discerns, "So what we have here is just the latest, and no doubt not the last, attempt to put God back into nature."

It is clear from the quotations, I think, that the reviewers are not rejecting design because there is scientific evidence against it, or because it violates some principle of logic. Rather, I believe they find design unacceptable because they are uncomfortable with the theo-logical ramifications of the theory. In his essay to the Pontifical Acad-emy of Science, Pope John Paul II noted that a theory of evolution has two parts, the mechanism and the philosophy attached to that mechanism. Putting it like that, however, makes it sound as if any philosophy can be mixed and matched with any mechanism. But the situation is not really that clear-cut. While Catholics and many other Christians could accommodate the mechanism of Darwin to their theology (with the reservation that the course of evolution is not truly random, but foreordained by God), materialists *require* some-thing like Darwinism because, ultimately, materialism says that life and intelligence had to arise unaided from brute matter.

A theory of intelligent design, however, holds implicitly that there is a designer capable of planning and executing the phenomenal intricacies of life on earth. Although there are, at least in theory, some exotic candidates for the role of designer that might be compatible with materialist philosophy (such as space aliens or time travelers), few people will be convinced by these and will conclude that the designer is beyond nature. Many scientists are unable or unwilling to accept such a designer because that goes against their prior commitment to materialism, or at least to a functional materialism in the course of their work.

Nonetheless, I remain optimistic that the scientific community will eventually accept intelligent design, even if the acceptance is discreet and muted. The reason for optimism is the advance of science itself, which almost every day discovers new intricacies in nature, fresh reasons for recognizing the design inherent in life and the universe.

8

WORD GAMES

DNA, Design, and Intelligence

STEPHEN C. MEYER

Stephen Meyer, Ph.D. (history and philosophy of science, Cambridge University), is associate professor of philosophy at Whitworth College, senior research fellow at the Discovery Institute (Seattle), and director of the Discovery Institute's Center for the Renewal of Science and Culture. He has contributed to several books and anthologies, including The History of Science and Religion in the Western Tradition,[1] *Darwinism: Science or Philosophy,[2] Of* Pandas and People,[3] The Creation Hypothesis,[4] *and* Facets of Faith and Science.[5]

> Einstein said, "God does not play dice."
> He was right. God plays scrabble.
>
> Philip Gold

*S*ince the late nineteenth century most biologists have rejected the idea that living organisms display evidence of intelligent design. While many acknowledge the *appearance* of design in biological systems, they insist that Darwinism, or neo-Darwinism, explains how this appearance arose naturalistically—that is, without invoking a directing intelligence or agency. Following Darwin, modern neo-Darwinists generally accept that natural selection acting on random variation can explain the appearance of design in living organisms.

102

As evolutionary biologist Francisco Ayala has explained, "The functional design of organisms and their features would ... seem to argue for the existence of a designer. It was Darwin's greatest accomplishment [however] to show that the directive organization of living beings can be explained as the result of a natural process, natural selection, without any need to resort to a Creator or other external agent."

Yet, however one assesses the explanatory power of Darwinism (or modern neo-Darwinism), the appearance of design in at least one important domain of biology cannot be so easily dismissed. During the last half of the twentieth century, advances in molecular biology and biochemistry have revolutionized our understanding of the miniature world within the cell. Research has revealed that cells—the fundamental units of life—store, transmit, and edit information and use that information to regulate their most fundamental metabolic processes. Far from characterizing cells as simple "homogeneous globules of plasm" as did Ernst Haeckel and other nineteenth-century biologists, biologists now describe cells as, among other things, "distributive real time computers" or complex information-processing systems. Recently, for example, a special issue of the prestigious journal *Cell*[6] was dedicated entirely to the topic of "macromolecular machines."

Darwin, of course, neither knew about these intricacies nor sought to explain their origin. Instead, his theory of biological evolution sought to explain how life could have grown gradually more complex *starting* from "one or a few simple forms." Strictly speaking, therefore, those who insist that the purely naturalistic Darwinian mechanism can explain the appearance of design in biology overstate their case. The complexities within the microcosm of the cell beg for some kind of explanation. Yet they lie beyond the purview of strictly biological evolutionary theory, which assumes, rather than explains, the existence of the first life and the information it required.

Explaining Life's Origin in Materialistic Terms

During the 1870s and 1880s scientists assumed that devising an explanation for the origin of life would be fairly easy. For one thing, they assumed that life was essentially a rather simple substance called protoplasm that could be easily constructed by combining and recombining simple chemicals such as carbon dioxide, oxygen, and nitrogen. Early theories of life's origin reflected this view. Haeckel likened cell "autogeny," as he called it, to the process of inorganic crystallization. Haeckel's English counterpart, T. H. Huxley, proposed a simple two-step method of chemical recombination to explain the origin of the first cell. Just as salt could be produced spontaneously by adding sodium to chloride, so, thought Haeckel and Huxley, could a living cell be produced by adding several chemical constituents together and then allowing spontaneous chemical reactions to produce the simple protoplasmic substance that they assumed to be the essence of life.

During the 1920s and 1930s a more sophisticated version of this so-called chemical evolutionary theory was proposed by a Russian biochemist named Alexander I. Oparin. Oparin had a much more accurate understanding of the complexity of cellular metabolism, but neither he, nor anyone else in the 1930s, fully appreciated the complexity of the molecules, such as protein and DNA, that make life possible. Oparin, like his nineteenth-century predecessors, suggested that life could have first evolved as the result of a series of chemical reactions. Unlike his predecessors, however, he envisioned that this process of chemical evolution would involve many more chemical transformations and reactions, and many hundreds of millions (or even billions) of years.

The first experimental support for Oparin's hypothesis came in December of 1952. While doing graduate work under Harold Urey at the University of Chicago, Stanley Miller circulated a gaseous mixture of methane, ammonia, water vapor, and hydrogen through a glass vessel containing an electrical discharge chamber. Miller sent a high voltage charge of electricity into the chamber via tungsten filaments in an attempt to simulate the effects of ultraviolet light on prebiotic atmospheric gases. After two days, Miller found a small (2 percent) yield of amino acids in the U-shaped water trap he used to collect reaction products at the bottom of the vessel.

Miller's success in producing biologically relevant "building blocks" under ostensibly prebiotic conditions was heralded as a great breakthrough. His experiment seemed to provide experimental support for Oparin's chemical evolutionary theory by showing that an important step in Oparin's scenario—the production of biological building blocks from simpler atmospheric gases—was possible on the early Earth.

104

Miller's experimental results also received widespread press coverage in popular publications such as *Time* magazine and gave Oparin's model the status of textbook orthodoxy almost overnight. Thanks largely to Miller's experimental work, chemical evolution is now routinely presented in both high school and college biology textbooks as the accepted scientific explanation for the origin of life.

Yet as we shall see, chemical evolutionary theory is now known to be riddled with difficulties, and Miller's work is understood by the origin-of-life research community itself to have little if any relevance to explaining how amino acids—let alone proteins or living cells—actually could have arisen on the early Earth.

To understand today's growing crisis in chemical evolutionary theory, this chapter will focus on the two most severe difficulties confronting it: the problem of hostile prebiotic conditions and the problem posed by the complexity of the cell and its components.

Hostile Prebiotic Conditions

When Stanley Miller conducted his experiment simulating the production of amino acids on the early Earth, he presupposed that the Earth's atmosphere was composed of a mixture of what chemists call reducing gases, such as methane, ammonia, and hydrogen. He also assumed that the Earth's atmosphere contained virtually no free oxygen. In the years following Miller's experiment, however, new geochemical evidence made it clear that the assumptions that Oparin and Miller had made about the early atmosphere could not be justified.

Instead, evidence strongly suggested that neutral gases—not methane, ammonia, and hydrogen—predominated in the early atmosphere. Moreover, a number of geochemical studies showed that significant amounts of free oxygen were also present even before the advent of plant life, probably as the result of volcanic outgassing and the photodissociation of water vapor.

In a chemically neutral atmosphere, reactions among atmospheric gases will not take place readily. Moreover, even a small amount of atmospheric oxygen will quench the production of biological building blocks and cause any biomolecules otherwise present to degrade rapidly.

As had been well known even before Miller's experiment, amino acids will form readily in an appropriate mixture of reducing gases.

What made Miller's experiment significant was not the production of amino acids *per se,* but the production of amino acids from presumably plausible prebiotic conditions. As Miller himself stated, "In this apparatus an attempt was made to duplicate a primitive atmosphere of the earth, and not to obtain the optimum conditions for the formation of amino acids." Now, however, the situation has changed. The only reason to continue assuming the existence of a chemically reducing, prebiotic atmosphere is that chemical evolutionary theory requires it.

Ironically, even if we assume for the moment that the reducing gases used by Stanley Miller do actually simulate conditions on the early Earth, his experiments inadvertently demonstrated the necessity of intelligent agency. Even successful simulation experiments require the intervention of the experimenters to prevent what are known as "interfering cross-reactions" and other chemically destructive processes. Without human intervention, Miller-type experiments invariably produce nonbiological substances that degrade amino acids into nonbiologically relevant compounds.

Experimenters prevent this by removing chemical products that induce undesirable cross-reactions. They employ other "unnatural" interventions as well. Simulation experimenters have typically used only short wavelength light, rather than both short and long wavelength ultraviolet light, which would be present in any realistic atmosphere. Why? The presence of the long-wavelength UV light quickly degrades amino acids.

Such manipulations constitute what chemist Michael Polanyi called a "profoundly informative intervention." They seem to "simulate," if anything, the need for an intelligent agent to overcome the randomizing influences of natural chemical processes.

Sequence Specificity in Proteins

Yet a more fundamental problem remains for all chemical evolutionary scenarios. Even if it could be demonstrated that the building blocks of essential molecules could arise in realistic prebiotic conditions, the problem of assembling those building blocks into functioning proteins or DNA chains would remain.

In the early 1950s, the molecular biologist Fred Sanger determined the structure of the protein molecule insulin. Sanger's work made clear for the first time that each protein found in the cell comprises a long and definitely arranged sequence of amino acids. The amino acids in protein molecules are linked together to form a chain, rather like individual railroad cars comprising a long train. Moreover, the function of all such proteins (whether as enzymes or as structural components in the cell) depends upon the specific sequencing of the individual amino acids, just as the meaning of an English text depends upon the sequential arrangement of the letters. The various chemical interactions between amino acids in any given chain determine a complex three-dimensional shape or topography that the amino acid chain adopts. This usually highly complex shape in turn determines what function, if any, the amino acid chain can perform within the cell. For a functioning protein, its three-dimensional shape gives it a hand-in-glove fit with other complex molecules in the cell, enabling it to catalyze specific chemical reactions or to build specific structures within the cell.

The discovery of the complexity and specificity of protein molecules has raised serious difficulties for chemical evolutionary theory, even if an abundant supply of amino acids is granted for the sake of argument. Amino acids alone do not make proteins, any more than letters alone make words, sentences, or poetry. In both cases, the sequencing of the constituent parts determines the function (or lack of function) of the whole. In the case of human languages, the sequencing of letters and words is obviously performed by intelligent human agents. In the cell, the sequencing of amino acids is directed by the information—the set of biochemical instructions—encoded on the DNA molecule.

Sequence Specificity in DNA

During the 1950s and 1960s, at roughly the same time molecular biologists began to determine the structure and function of many proteins, scientists were able to explicate the structure and function of DNA, the molecule of heredity. After James Watson and Francis Crick elucidated the structure of DNA in 1953, molecular biologists soon discovered how DNA directs the process of protein synthesis within the cell. They discovered that the specificity of amino acids in proteins derives from a prior specificity within the DNA molecule—from information on the DNA molecule stored as millions of specifically arranged chemicals called nucleotides or bases along the spine of the DNA's helical strands. Chemists represent the four nucleotides with the letters A, T, G, and C (for adenine, thymine, guanine, and cytosine).

As it turns out, specific regions of the DNA molecule called coding regions have the same property of "sequence specificity" or "specified complexity" that characterizes written codes, linguistic texts, and protein molecules. Just as the letters in the alphabet of a written language may convey a particular message depending on their arrangement, so too do the sequences of nucleotide bases (the As, Ts, Gs, and Cs) inscribed along the spine of a DNA molecule convey a precise set of instructions for building proteins within the cell. The nucleotide bases in DNA function in much the same way as symbols in a machine code or alphabetic characters in a book.

In each case, the arrangement of the characters determines the function of the sequence as a whole. As Richard Dawkins has noted, "The machine code of the genes is uncannily computer-like." Or as Bill Gates has noted, "DNA is like a computer program, but far, far more advanced than any software we've ever created." In the case of a computer code, the specific arrangement of just two symbols (0 and 1) suffices to carry information. In the case of an English text, the twenty-six letters of the alphabet do the job. In the case of DNA, the complex but precise sequencing of the four nucleotide bases (A, T, G, and C) stores and transmits the information necessary to build proteins. Thus, the sequence specificity of proteins derives from a prior sequence specificity—from the *information*—encoded in DNA.

The elucidation of DNA's information-bearing properties raised the question of the ultimate origin of the information in both DNA and proteins. Indeed, many scientists now refer to the information problem as the "Holy Grail" of origin-of-life biology. As Bernd-Olaf Küppers recently stated, "the problem of the origin of life is clearly basically equivalent to the problem of the origin of biological information." Since the 1950s, three broad types of naturalistic explanation have been proposed by scientists to explain the origin of information: chance, prebiotic natural selection, and chemical necessity.

108

Beyond the Reach of Chance

While many outside origin-of-life biology may still invoke chance as a causal explanation for the origin of biological information, few serious researchers still do. Since molecular biologists began to appreciate the sequence specificity of proteins and nucleic acids in the 1950s and 1960s, many calculations have been made to determine the probability of formulating functional proteins and nucleic acids at random. Even assuming extremely favorable prebiotic conditions (whether realistic or not) and theoretically maximal reaction rates, such calculations have invariably shown that the probability of obtaining functionally sequenced biomacromolecules at random is, in Ilya Prigogine's words, "vanishingly small . . . even on the scale of . . . billions of years." As A. Graham Cairns-Smith wrote:

> Blind chance . . . is very limited. [Blind chance can produce] low levels of cooperation . . . exceedingly easily (the equivalent of letters and small words), but it becomes very quickly incompetent as the amount of organization increases. Very soon indeed long waiting periods and massive material resources become irrelevant.[7]

Consider the probabilistic hurdles that must be overcome to construct even one short protein molecule of about 100 amino acids in length. First, all amino acids must form a chemical bond known as a peptide bond so as to join with other amino acids in the protein chain. Yet in nature many other types of chemical bonds are possible between amino acids. The probability of building a chain of 100 amino acids in which all linkages involve peptide bonds is roughly 1 chance in 10^{30}.

Second, in nature every amino acid has a distinct mirror image of itself, one left-handed version or L-form and one right-handed version or D-form. These mirror-image forms are called optical isomers. Functioning proteins tolerate only left-handed amino acids, yet the right-handed and left-handed isomers occur in nature with roughly equal frequency. Taking this into consideration compounds the improbability of attaining a biologically functioning protein. The probability of attaining at random only L-amino acids in a hypothetical peptide chain 100 amino acids long is $(\frac{1}{2})^{100}$ or again roughly 1 chance in 10^{30}.

Third and most important of all, functioning proteins must have amino acids that link up in a specific sequential arrangement, just like the letters in a meaningful sentence. Because there are 20 biologically occurring amino acids, the probability of getting a specific amino acid at a given site is $\frac{1}{20}$. Even if we assume that some sites along the chain will tolerate several amino acids (using the variances determined by biochemist Robert Sauer of MIT), we find that the probability of achieving a functional sequence of amino acids in several functioning proteins at random is still "vanishingly small," roughly 1 chance in 10^{65}—an astronomically large number—for a protein only one hundred amino acids in length. (Actually the probability is even lower because there are many nonproteinous amino acids in nature that we have not accounted for in this calculation.)

If one also factors in the probability of attaining proper bonding and optical isomers, the probability of constructing a rather short, functional protein at random becomes so small as to be effectively zero (no more than 1 chance in 10^{125}), even given our multi-billion-year-old universe. Consider further that equally severe probabilistic difficulties attend the random assembly of functional DNA. Moreover, a minimally complex cell requires not 1, but at least 100 complex proteins (and many other biomolecular components such as DNA and RNA) all functioning in close coordination. For this reason, quantitative assessments of cellular complexity have simply reinforced an opinion that has prevailed since the mid-1960s within origin-of-life biology: chance is not an adequate explanation for the origin of biological complexity and specificity.

Natural Selection a Dead End

At nearly the same time that many researchers became disenchanted with chance explanations, theories of prebiotic natural selection also fell out of favor. Such theories allegedly overcame the difficulties of pure chance by providing a mechanism by which complexity-increasing events in the cell might be preserved and selected. Yet these theories shared many of the difficulties that afflict purely chance-based theories.

Natural selection presupposes a preexisting mechanism of self-replication. Yet, self-replication in all extant cells depends upon functional (and, therefore, to a high degree sequence-specific) proteins and nucleic acids. The origin of these molecules is precisely what

110

Oparin needed to explain. Thus, many rejected his postulation of prebiotic natural selection as begging the question. As the evolutionary biologist Dobzhansky would insist, "prebiological natural selection is a contradiction in terms."

Further, natural selection can only select what chance has first produced, and chance, at least in a prebiotic setting, seems an implausible agent for producing the information present in even a single functioning protein or DNA molecule. As Christian de Duve has explained, theories of prebiotic natural selection "need information which implies they have to presuppose what is to be explained in the first place." For this reason, most scientists now dismiss appeals to prebiotic natural selection as essentially indistinguishable from appeals to chance.

Self-Organization

Because of these difficulties, many origin-of-life theorists after the mid-1960s attempted to address the problem of the origin of biological information in a completely new way. Rather than invoking prebiotic natural selection or "frozen accidents," many theorists suggested that the laws of nature and chemical attraction may themselves be responsible for the information in DNA and proteins. Some have suggested that simple chemicals might possess "self-ordering properties" capable of organizing the constituent parts of proteins, DNA, and RNA into the specific arrangements they now possess. Just as electrostatic forces draw sodium (Na+) and chloride ions (Cl-) together into a highly ordered pattern within a crystal of salt (NaCl), so too might amino acids with special affinities for each other arrange themselves to form proteins. Kenyon and Steinman developed this idea in a book entitled *Biochemical Predestination*[8] in 1969.

In 1977, Prigogine and Nicolis proposed another theory of self-organization based on their observation that open systems driven far from equilibrium often display self-ordering tendencies. For example, gravitational energy will produce highly ordered vortices in a draining bathtub, and thermal energy flowing through a heat sink will generate distinctive convection currents or "spiral wave activity."

For many current origin-of-life scientists, self-organizational models now seem to offer the most promising approach to explaining the origin of biological information. Nevertheless, critics have called into question both the plausibility and the relevance of self-

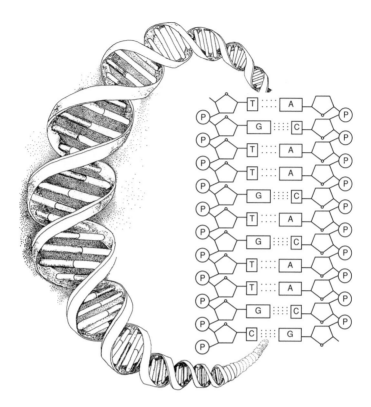

The bonding relationships between the chemical constituents of the DNA molecule. Sugars (designated by the pentagons) and phosphates (designated by the circled Ps) are chemically linked. Nucleotide bases (the As, Ts, Gs, and Cs) are bonded to the sugar-phosphate backbones. Nucleotide bases are linked by hydrogen bonds (designated by dotted double or triple lines) across the double helix. But no chemical bonds exist between the nucleotide bases along the message-bearing spine of the helix, demonstrating that physical and chemical forces are not responsible for the specific sequencing in the molecule.

organizational models. Ironically, perhaps the most prominent early advocate of self-organization, Dean Kenyon, has now explicitly repudiated such theories as both incompatible with empirical findings and theoretically incoherent.

The empirical difficulties attendant on self-organizational scenarios can be illustrated by examining a DNA molecule. The diagram above shows that the structure of DNA depends upon several chemical bonds. There are bonds, for example, between the sugar and the

112

phosphate molecules that form the two twisting backbones of the DNA molecule. There are bonds fixing individual (nucleotide) bases to the sugar-phosphate backbones on each side of the molecule. Yet notice that there are no chemical bonds between the bases that run along the spine of the helix. Yet it is precisely along this axis of the molecule that the genetic instructions in DNA are encoded.

Further, just as magnetic letters can be combined and recombined in any way to form various sequences on a metal surface, so too can each of the four bases A, T, G, and C attach to any site on the DNA backbone with equal facility, making all sequences equally probable (or improbable). The same type of chemical bond occurs between the bases and the backbone regardless of which base attaches. All four bases are acceptable; none is preferred. In other words, *differential* bonding affinities do not account for the sequencing of the bases. Because these same facts hold for RNA molecules, researchers who speculate that life began in an "RNA world" have also failed to solve the sequencing problem—i.e., the problem of explaining how information present in all functioning RNA molecules could have arisen in the first place.

For those who want to explain the origin of life as the result of self-organizing properties intrinsic to the material constituents of living systems, these rather elementary facts of molecular biology have devastating implications. The most logical place to look for self-organizing properties to explain the origin of genetic information is in the constituent parts of the molecules carrying that information. But biochemistry and molecular biology make it clear that the forces of attraction between the constituents in DNA, RNA, and protein do not explain the sequence specificity of these large information-bearing biomolecules.

Significantly, information theorists insist that there is a good reason for this. If chemical affinities between the constituents in the DNA message text determined the arrangement of the text, such affinities would dramatically diminish the capacity of DNA to carry information. Consider what would happen if the individual nucleotides (A, T, G, and C) in a DNA molecule *did* interact by *chemical* necessity with each other. Every time adenine (A) occurred in a growing genetic sequence, it would likely drag thymine (T) along with it. Every time cytosine (C) appeared, guanine (G) would follow. As a result, the DNA message text would be peppered with repeating sequences of As followed by Ts and Cs followed by Gs.

Rather than having a genetic molecule capable of unlimited novelty, with all the unpredictable and aperiodic sequences that characterize informative texts, we would have a highly repetitive text awash in redundant sequences—similar to what happens in crystals. Indeed, in a crystal the forces of mutual chemical attraction do completely explain the sequential ordering of the constituent parts, and consequently crystals cannot convey novel information. Sequencing in crystals is repetitive and highly ordered, but not informative. Once one has seen "Na" followed by "Cl" in a crystal of salt, for example, one has seen the extent of the sequencing possible.

Bonding affinities, to the extent they exist, undercut the maximization of information. They cannot, therefore, be used to explain the origin of information. Affinities create mantras, not messages.

The tendency to confuse the qualitative distinction between "order" and "information" has characterized self-organizational research efforts and calls into question the relevance of such work to the origin of life. Self-organizational theorists explain well what doesn't need explaining. What needs explaining is not the origin of order (whether in the form of crystals, swirling tornadoes, or the eyes of hurricanes), but the origin of *information*—the highly improbable, aperiodic, and yet specified sequences that make biological function possible.

Information, Design, and Intelligence

To see the distinction between order and information, compare the sequence "ABABABABABABAB" to the sequence "Help! Our neighbor's house is on fire!" The first sequence is repetitive and ordered, but not complex or informative. Systems that are characterized by both specificity and complexity (what information theorists call specified complexity) have information content. Since such systems have the qualitative feature of aperiodicity or complexity, they are qualitatively distinguishable from systems characterized by simple periodic order. Thus, attempts to explain the origin of order have no relevance to discussions of the origin of information content.

Significantly, the nucleotide sequences in the coding regions of DNA have, by all accounts, a high information content—that is, they are both highly specified and complex, just like meaningful

English sentences or functional lines of code in computer software. Yet the information contained in an English sentence or computer software does not derive from the chemistry of the ink or the physics of magnetism, but from a source extrinsic to physics and chemistry altogether. Indeed, in both cases, the message transcends the properties of the medium.

The information in DNA also transcends the properties of its material medium. Because chemical bonds do not determine the arrangement of nucleotide bases, the nucleotides can assume a vast array of possible sequences and thereby express many different biochemical messages.

If the properties of matter (i.e., the medium) do not suffice to explain the origin of information, what does? Our experience with information-intensive systems (especially codes and languages) indicates that such systems always come from an intelligent source—i.e., from mental or personal agents, not chance or material necessity.

This generalization about the cause of information has, ironically, received confirmation from origin-of-life research itself. During the last forty years, every naturalistic model proposed has failed to explain the origin of information—the great stumbling block for materialistic scenarios. Thus, mind or intelligence or what philosophers call "agent causation" now stands as the only cause known to be capable of creating an information-rich system, including the coding regions of DNA, functional proteins, and the cell as a whole.

Because mind or intelligent design is a necessary cause of an informative system, one can detect the past action of an intelligent cause from the presence of an information-intensive effect, even if the cause itself cannot be directly observed. Since information requires an intelligent source, the flowers spelling "Welcome to Victoria" in the gardens of Victoria Harbor in Canada lead visitors to infer the activity of intelligent agents even if they did not see the flowers planted and arranged.

Scientists in many fields now recognize the connection between intelligence and information and make inferences accordingly. Archaeologists assume a mind produced the inscriptions on the Rosetta Stone. SETI's search for extraterrestrial intelligence presupposes that the presence of information imbedded in electromagnetic signals from space would indicate an intelligent source. As yet, radio astronomers have not found information-bearing signals coming from space. But molecular biologists, looking closer to home, have discovered information in the cell. Consequently, DNA justifies making what probability theorist William Dembski calls "the design inference."

God of the Gaps?

Of course, many scientists have argued that to infer design gives up on science. They say that inferring design constitutes an argument from scientific ignorance—a "God of the gaps" fallacy. Since science doesn't yet know how biological information could have arisen, design theorists invoke a mysterious notion—intelligent design—to fill a gap in scientific knowledge.

Yet design theorists do not infer design just because natural processes cannot explain the origin of biological systems, but because these systems manifest the distinctive hallmarks of intelligently designed systems—that is, they possess features that in any other realm of experience would trigger the recognition of an intelligent cause. For example, Michael Behe has inferred design not only because the gradualistic mechanism of natural selection cannot produce irreducibly complex systems, but also because in our experience irreducible complexity is a feature of systems known to have been intelligently designed. That is, whenever we see systems that have the feature of irreducible complexity and we know the causal story about how such systems originated, invariably intelligent design played a role in the origin of such systems. Thus, Behe infers intelligent design as the best explanation for the origin of irreducible complexity in cellular molecular motors, for example, based upon what we *know*, not what we do not know, about the causal powers of nature and intelligent agents, respectively.

Similarly, the specified complexity or information content of DNA and proteins implicates a prior intelligent cause, because specified complexity and high information content constitute a distinctive hallmark (or signature) of intelligence. Indeed, in all cases where we know the causal origin of high information content or specified complexity, experience has shown that intelligent design played a causal role. Thus, when we encounter such information in the biomacromolecules necessary to life, we may infer—based upon our *knowledge* of established cause-and-effect relationships—that an intelligent cause operated in the past to produce the information necessary for the origin of life.

Design theorists infer a prior intelligent cause based upon present knowledge of cause-and-effect relationships. Inferences to design thus employ the standard uniformitarian method of reasoning used in all historical sciences, many of which routinely detect intelligent causes. We would not say, for example, that an archeologist had committed a "scribe of the gaps" fallacy simply because he inferred that an intelligent agent had produced an ancient hieroglyphic inscription. Instead, we recognize that the archeologist has made an inference based upon the *presence* of a feature (namely, high information content) that invariably implicates an intelligent cause, not (solely) upon the *absence* of evidence for a suitably efficacious natural cause.

Intelligent agents have unique causal powers that nature does not. When we observe effects that we know only agents can produce, we rightly infer the presence of a prior intelligence even if we did not observe the action of the particular agent responsible. Since DNA displays an effect (namely, information content) that in our experience only agents can produce, intelligent design (and not apparent design) stands as the best explanation for the information content in DNA.

9

MAKING SENSE
OF BIOLOGY

The Evidence for Development by Design

JONATHAN WELLS

Jonathan Wells, Ph.D. (religious studies, Yale University, and developmental biology, University of California at Berkeley), is a postdoctoral research biologist at the University of California at Berkeley, and a fellow of the Discovery Institute's Center for the Renewal of Science and Culture. He is the author of Charles Hodge's Critique of Darwinism[1] *and* Icons of Evolution.[2]

A developing embryo is an amazing thing to watch. First, a single cell divides into many cells; then, with uncanny precision, those cells rearrange themselves into the basic shape of the animal, forming a gut in the process; and finally, small groups of cells form specific structures such as eyes and limbs.

As a developmental biologist, I study embryos primarily by perturbing their development. I am constantly impressed by their resilience. Despite my interference, a surprising number of them develop to adulthood. Remarkably, although my interference may introduce various deformities, the basic endpoint of development never changes. If they survive, fruit fly eggs always become fruit flies, frog eggs always become frogs, and mouse eggs always become mice. Not even the species changes.

118

So every embryo is somehow programmed to develop into a particular species of animal. For centuries, embryologists have been trying to understand the nature of this developmental program. Now, finally, many think they have succeeded.

In this essay I will describe the understanding of genetic programs that is currently most popular, and present evidence against it. I will then argue that its popularity depends largely on its logical connection with the neo-Darwinian theory of evolution. Neo-Darwinism itself, however, is based less on scientific evidence than on a philosophically motivated desire to exclude the idea that living things are designed. I will conclude by arguing that this exclusion is unwarranted and that acknowledging the possibility of design would contribute to future progress in the study of embryonic development.

Genetic Programs: Theory and Counterevidence

In 1865, Gregor Mendel attributed patterns of inheritance in pea plants to factors that were later called genes. When Watson and Crick discovered the structure of DNA in the 1950s, Mendelian genetics found a molecular basis, and in 1973 Sydney Brenner suggested that embryonic development might be directed by a genetic "program" encoded in DNA.

During the 1970s and 1980s, developmental geneticists made significant progress in understanding the role of genes in fruit fly development. Some of the genes they identified are involved in patterning the embryo from head to rear, others are involved in patterning it from back to front, and still others are involved in specifying the identities of particular cells. Since 1990, developmental geneticists have made similar progress with other organisms, including mammals, and the idea that development is directed by a genetic program is now widely held.

Widely, but not universally. According to a small but growing number of biologists, there is considerable evidence that genes do *not* control development. For example, when an egg's genes are removed and replaced with genes from another type of animal, development follows the pattern of the original egg until the embryo dies from lack of the right proteins. The "Jurassic Park" approach of putting dinosaur DNA into ostrich eggs to produce a *Tyrannosaurus rex* makes exciting fiction, but it ignores scientific fact.

119

What about mutating the DNA instead of replacing it completely? Using a technique called "saturation mutagenesis," biologists have found that mutations in developmental genes often lead to death or deformity, but they never produce changes that benefit the organism. Furthermore, DNA mutations never alter the endpoint of embryonic development; they cannot even change the species. An embryo needs the right genes to make new proteins, and its development suffers without them; but being dependent on genes is not the same as being controlled by them. A house under construction needs suitable building materials, but they do not determine its floor plan.

If DNA were in control of development, then I should be able to produce a replica of myself by putting my DNA in a human egg that has had its own DNA removed. This is the reasoning behind the uproar over cloning. But not even identical twins are replicas of each other; they frequently differ somewhat in physical characteristics, and they always differ—sometimes dramatically—in temperament and interests. Yet identical twins share not only the same DNA, but also the same egg cell and (usually) the same womb. Even the imperfect similarity exhibited by identical twins requires more than the same DNA.

Ironically, the very success of DNA-transfer experiments provides evidence against genetic programs since it shows that adult cells contain the same DNA as a fertilized egg. But the cells of an adult animal differ markedly from each other in form and function. If they have the same DNA, why are they so different? Part of the answer is that each cell type utilizes only a portion of its genetic repertoire, with factors outside of the DNA turning on the appropriate genes. But if development requires that DNA be controlled by factors outside of itself, then DNA does not control development.

Recently, biologists have found that the genes that seem to be most important in development are remarkably similar in many different types of animals, from worms to fruit flies to mammals. Initially, this was regarded as evidence for genetic programs, but biologists are now realizing that it actually constitutes a paradox: If genes control development, why do similar genes produce such different animals? Why does a caterpillar turn into a butterfly instead of a barracuda?

If DNA does not control development, what does? Actually, there is good evidence for the involvement of at least two other factors in the egg—the cytoskeleton and the membrane. Every animal cell contains a network of microscopic fibers called a cytoskeleton. These fibers include microtubules, which are known to be involved in pat-

120

DNA Is Not Destiny

Evidence That Genetic Programs Do Not Control Development

1. Placing foreign DNA into an egg does not change the species of the egg or embryo.
2. DNA mutations can interfere with development, but they never alter its endpoint.
3. Different cell types arise in the same animal even though all of them contain the same DNA.
4. Similar developmental genes are found in animals as different as worms, flies, and mammals.
5. Eggs contain several structures (such as microtubule arrays and membrane patterns) that are known to exercise control of development independently of the DNA.

Evidence Against Neo-Darwinism

1. Embryonic development is not controlled by a genetic program.
2. Mutations do not produce the sorts of changes needed for evolution.
3. Except at the level of antibiotic and insecticide resistance, there are no good examples of evolution due to changes in gene frequencies produced by natural selection.

terning embryos. For example, one of the gene products involved in head-to-rear patterning of fruit fly embryos is delivered to its proper location by microtubules. If the microtubules are experimentally disrupted, the gene product doesn't reach its proper destination and the embryo is grossly deformed.

Microtubules consist of many identical protein subunits, and each subunit is produced according to a template in the organism's DNA. But what matters in development is the shape and location of microtubule arrays, and the shape and location of a microtubule array is not determined by its subunits any more than the shape and location of a house is determined by its bricks. Instead, microtubule arrays are formed by mysterious organelles called centrosomes, which are inherited independently of an organism's DNA. Centrosomes play a central role in development. A frog egg can be induced to develop into a parthenogenetic frog merely by injecting a sperm centrosome; no sperm DNA is needed.

Another nongenetic factor involved in development is the membrane pattern of the egg cell. Cell membranes are not merely featureless bags, but highly complex structures. For example, a membrane contains specialized channels that pump molecules in and out of the cell, enabling it to control its interactions with the external environment. An egg cell membrane also contains "targets" that ensure that molecules synthesized in the nucleus reach their proper destinations in the embryo. The gene product mentioned above that is involved in head-to-rear patterning of fruit fly embryos and that depends on microtubules to deliver it to its proper location also needs a target molecule to keep it in place after it arrives. The target is already there, embedded in the membrane.

Experiments with single-celled animals show that membrane patterns are determined by preexisting membranes, not by DNA. Like microtubule subunits, the proteins embedded in a membrane are produced according to templates in the organism's DNA; but like the form and location of microtubule arrays, the patterns of those embedded proteins are inherited independently of the organism's DNA. So the control exercised by microtubule arrays and membrane patterns over embryonic development is not encoded in DNA sequences.

This does not mean that we now understand developmental programs. Far from it! But it is quite clear that they cannot be reduced to *genetic* programs, written in the language of DNA sequences. It would be more accurate to say that a developmental program is written into the structure of the entire fertilized egg—including its DNA, microtubule arrays, and membrane patterns—in a language of which we are still largely ignorant.

Why, then, does the notion of genetic programs continue to be so popular? The answer, to a great extent, depends on its logical connection with neo-Darwinian evolution.

Genetic Programs and Neo-Darwinism

Genetic programs are a corollary of the 1930s synthesis of Mendelian genetics and Darwinian evolution, now commonly called "neo-

122

Darwinism."According to neo-Darwinism, genetic mutations provide the raw materials for evolution, and natural selection modifies organisms through changes in gene frequencies. Development is what turns a single cell into a worm instead of a mouse; so if evolution can change worms into mice by modifying their genes, then it must do so by modifying genes that control development. Conversely, if development is controlled by something other than genes, then evolution must be due to something other than genetic mutations and changes in gene frequencies; if the notion of genetic programs is false, so is neo-Darwinism. Thus neo-Darwinism logically entails genetic programs.

Many biologists assume that there is a lot of evidence for neo-Darwinism. If this were so, then belief in genetic programs might be justified in spite of the evidence described above. The evidence for neo-Darwinism, however, turns out to be surprisingly thin.

Mutations are supposed to provide raw materials for evolution, but they can do this only if they benefit the organism, and mutations in developmental genes are always harmful. In fact, the only DNA mutations that are known to be beneficial are those that affect immediate interactions between a mutant protein and other molecules. Such mutations can confer antibiotic and insecticide resistance, but they never lead to the sorts of changes that could account for evolution. DNA mutations cannot even change the species of an animal, much less change a fish into an amphibian or a dinosaur into a bird.

The evidence that evolution is due to changes in gene frequencies is likewise surprisingly thin. Outside of antibiotic and insecticide resistance, the best-studied example is industrial melanism in peppered moths. During the industrial revolution, dark (melanic) peppered moths became more common than light moths, until their proportion declined with the advent of pollution-control legislation. The rise and fall of melanism certainly involved changes in gene frequencies, and experiments in the 1950s seemed to show that these were a result of natural selection: when caged moths were released onto pollution-darkened tree trunks, birds preyed selectively on the more conspicuous light moths, and the proportion of melanic moths increased. In the 1970s, however, biologists noticed that the proportions of light and dark moths in the wild did not correlate with bark color, and in the 1980s they learned that peppered moths do not normally rest on tree trunks. So the evidence for natural selection has been discredited, and the relevance of industrial melanism to evolution is in doubt. Unfortunately, in other supposed cases of evolution by natural selection—such as beak differences in Darwin's finches and adaptive divergence in Hawaiian fruit flies—the genetic basis is unknown.

The flimsiness of the evidence for neo-Darwinism indicates that its popularity must be due to something other than empirical corroboration. That something is a philosophical exclusion of design.

The Bias Against Design

In 1802, William Paley published his *Natural Theology*.[3] Toward the beginning of that book he noted that someone crossing a heath and finding a stone could reasonably attribute its presence to purposeless natural causes. Yet upon finding a watch and discovering that its parts were put together for a purpose, one could conclude that the watch had been designed. By analogy, Paley argued, one could also conclude that living things are designed.

Charles Darwin argued that living things are more like a stone than a watch, and that features of living things that Paley attributed to design were actually due to the operation of purposeless natural causes. Initially, Darwin's theory of evolution by natural selection failed to convince biologists because it was unable to explain the origin and inheritance of variations. (Mendel's work did not become widely known until after 1900.) But Darwin's exclusion of design caught on and soon became part of the very definition of science. Nowadays, anyone naive enough to argue for design in a biology classroom is invariably told that the notion is unscientific. Design is ruled out *a priori*.

Once design is ruled out, neo-Darwinism wins by default. If we must explain the origin and diversification of living things without recourse to design, then mutation and natural selection may be the only way to go. As atheist Richard Dawkins wrote in *The Blind Watchmaker*, "Darwinism is the only known theory that is in principle *capable* of explaining certain aspects of life."[4] Thus neo-Darwinism survives even though the evidence fails to support it.

The exclusion of design, however, is a philosophical move rather than a scientific one. The truth is that design is not inherently unscientific. Inferences to design are a normal exercise of rational thought, and we rely on them in a wide variety of fields. All design inferences follow a logical pattern that William Dembski calls "the explanatory filter." When inquiring into the cause of any phenomenon, we first attempt to explain it in terms of natural regularities; if we cannot, we ask whether it could plausibly be attributed to chance; only if there is sufficient evidence to rule out regularity and chance do we then infer design. A detective uses similar criteria to determine whether a criminal act was intentional; but if the detective expects to convince a jury, the determination had better be based on evidence—i.e., it must be scientific. It is not arbitrary, nor is it a confession of ignorance.

Darwinists typically argue that design inferences are excluded from biology because science is limited to phenomena that can be studied empirically, and design in living things would imply a supernatural designer who is beyond the reach of empirical methods. But there is a difference between seeing evidence for design in living things and speculating on the nature of the designer. Darwinists tacitly acknowledge this difference when they use evidence to argue *against* design, and thereby treat design as an empirical question while denying the very existence of a designer. (The subtitle of Richard Dawkins's *The Blind Watchmaker* is *Why the Evidence of Evolution Reveals a Universe Without Design.*) Of course, if it is scientific to use evidence to argue *against* design, it is just as scientific to use evidence to argue *for* it.

What if living things really *are* designed? Someone who discovers a watch on the ground and wants to investigate its origin would be foolish to rule out design at the outset, even if there were no hope of ever locating the designer. In addition to being mistaken about the true nature of the watch, such a person might waste an entire lifetime trying to figure out how the watch assembled itself.

Development by Design

If living things are designed, then it is a mistake to assume that they are not. Research based on a mistaken assumption may turn up interesting facts (as developmental geneticists have done), but many of the questions it seeks to address can never be answered because they are misconceived. Until the mistaken assumption is revised to fit the facts, researchers will be looking in the wrong place.

Embryos certainly *appear* to be designed. Ironically, this would be true even if development were controlled by a genetic program. As

Follow the Evidence

Reasons to Infer Design

1. Individual protein-coding regions in DNA cannot result from natural regularities or chance, and show clear evidence of design.
2. Genetic programs would have to include more information than protein-coding regions, so the neo-Darwinian attempt to exclude design by insisting on genetic programs fails.
3. Developmental programs are much more complex than genetic programs, rendering a design inference even more compelling.

Design and Science

1. The neo-Darwinian exclusion of design is based not on evidence, but on metaphysical naturalism—the philosophical doctrine that nature is all there is.
2. This exclusion of design leads to the insistence that development is controlled by genetic programs.
3. This insistence on genetic programs misguides biological research by discounting evidence for non-genetic factors in development.
4. Acknowledging the possibility of design does not imply that biologists should focus on proving the existence of a designer; it merely encourages them to follow the evidence wherever it leads.

Stephen Meyer has shown, protein-coding regions in DNA cannot be explained on the basis of natural regularities such as the laws of chemistry, and it is utterly implausible to attribute such highly specified sequences to chance (see his essay in chapter 8 of this book). The explanatory filter shows that they justify an inference to design. If this is true of individual protein-coding regions, it would be even more true of genetic programs, since they would have to contain a great deal of additional information regulating the expression of protein-coding regions.

The fact that development is actually controlled by something more complex than a genetic program—perhaps by the structure

126

of the entire fertilized egg—renders design even more probable. This is not to say that an individual embryo develops by design; if we understood the developmental program in any particular case, we would presumably attribute the development of that individual to natural regularity. But the existence of the developmental program itself cannot be accounted for by natural regularities (other than to say it is inherited from the parents, which merely moves the problem back one generation). Furthermore, the integrated complexity of developmental programs cannot plausibly be attributed to chance. Unless one excludes design on philosophical grounds, one would be justified in concluding, at least tentatively, that developmental programs are designed.

Does this mean that biologists should devote their energies to proving the existence of a designer? I think not. It simply means that biologists should trust their common sense. In *The Blind Watchmaker*, Richard Dawkins defines biology as the study of complex things that *appear* to be designed, but aren't. Instead of arbitrarily dismissing design as mere appearance, however, biologists would be better off following the evidence wherever it leads.

One immediate benefit would be to liberate biologists from the compulsion to reduce development to genetic programs. Instead of trying to force nongenetic factors into the Procrustean bed of neo-Darwinism, biologists would be encouraged to study them in their own right. Such factors are already being studied, of course, but the present neo-Darwinian monopoly in biology ensures that they consistently take a back seat to DNA-centered research. Once that monopoly is broken, we may expect to see dramatic scientific progress in understanding how embryos develop.

In 1973 Theodosius Dobzhansky wrote in *The American Biology Teacher* that "nothing in biology makes sense except in the light of evolution."[5] (By "evolution," Dobzhansky meant neo-Darwinism.) The flimsiness of the evidence for neo-Darwinism, however, exposes Dobzhansky's statement as a philosophical credo, not a scientific inference. Insisting that everything in biology should be subordinated to a theory—especially a theory with as little empirical support and as much metaphysical baggage as neo-Darwinism—is the antithesis of good science. Dobzhansky was wrong. Nothing in biology makes sense except in the light of evidence—even if the evidence points to design.

10

UNFIT FOR SURVIVAL

The Fatal Flaws of Natural Selection

PAUL A. NELSON

Paul Nelson, Ph.D. (philosophy of biology, University of Chicago), is editor of the journal Origins & Design, *and a fellow of the Discovery Institute's Center for the Renewal of Science and Culture. His article "The Role of Theology in Current Evolutionary Reasoning,"[1] appeared in* Biology and Philosophy. *He is a member of the International Society for the History, Philosophy, and Social Studies of Biology.*

*I*magine a calm and sunny afternoon in early summer. You are sitting on a bench on the north bank of the Charles River, near the Massachusetts Institute of Technology (MIT). You have read today's *Boston Globe*, and now, gazing abstractedly across the river at the passing rowers and distant traffic, you would like nothing better than to drift off into a daydream. Unfortunately a pesky fly has come buzzing around the bench. Fortunately the *Globe* is available. You take aim, and the calm returns.

Only for a moment, however. Suddenly you are surrounded by a large group of panting, ashen-faced graduate students. They stare at you in horrified disbelief, while one of them bends to retrieve the remains of the fly from the grass. "Our *fly* . . . ," he stammers in a shocked tone. *"You smashed our fly!"*

All right, so you're not an entomologist. But (you say in your defense) does one fly more or less make any difference to the global flying insect census?

In fact, what you just unwittingly destroyed was no ordinary fly. It was the only working prototype of MIT's Artificial Insect Project. This ingenious construction of microscopic motors and sensors, synthetic tissues, and thousands of lines of complicated programming is—or, rather, *was*—a marvel of nanotechnology. Indeed, had you looked carefully before swinging that rolled-up newspaper, you might have noticed that this was no ordinary fly.

Its microscopic primary motors, although cooled by sophisticated, specially formulated lubricants, tended to overheat with long use. The synthetic membrane that formed its wings would tear at high speeds, causing aerodynamic instability; hence, the fly often crashed into things. Its complex behavioral software required frequent debugging. No one knew, in fact, if the artificial fly would work at all from one day to the next—unlike the ordinary flies that we destroy with hardly a second thought.

Indeed, the artificial fly was in all respects hopelessly the functional inferior of its natural counterpart. Any curiosity we might feel about the causal origins of the artificial fly should, therefore, be correspondingly stronger about the causal origins of its natural counterpart.

Consider another perspective. In the memorable thought experiment that opens his book *Chance and Necessity*,[2] Jacques Monod describes a machine sent to Earth from an agency of an alien civilization—"a Martian NASA," as he puts it. The machine has been programmed to detect evidence "of organized, artifact-producing activity," of the sort displayed (in our experience) only by "intelligent beings capable of projective activity." The machine lands in a forest and turns on its instruments—which come to focus on a hive of wild bees. Now the computer controlling the machine, Monod argues,

> cannot take [the bees] for anything but artificial, highly elaborated objects. . . . [E]xamining bee after bee the computer will note the extreme complexity of their structure (the number and position of abdominal hairs, for example, or the ribbing of the wings) is reproduced with extraordinary fidelity from one individual bee to the next. Powerful evidence, is it not, that these creatures are the products of a deliberate, constructive, and highly sophisticated order of activity? Upon the basis of such conclusive documentation, the machine would be bound to signal to the officials of the

Martian NASA its discovery, upon Earth, of an industry compared with which their own would probably seem primitive."[3]

"Living creatures," Monod argues, "are strange objects." They exhibit distinctive properties, such as goal-directedness, which we ordinarily associate with designed objects, such as our own artifacts. Moreover, in their "construction" (i.e., in their reproduction and development), living things call on large stores of information "whose source," Monod continues, "has still to be identified: for all ... information presuppose[s] a source."[4]

Darwin's Simple Design

The notions of an artificial fly, or of Monod's Martian Search for Extraterrestrial Intelligence (SETI) machine, are of course in one sense fanciful conceits. Yet in another sense, these thought experiments throw vivid light on a profoundly motivating intuition. Suppose something like an artificial fly did land on one's windowsill? The intense curiosity we would undoubtedly feel about its design and construction—which may be captured by the question, Who or what built this?—arises from our uniform experience that objects having the specified complexity of an artificial fly require, as Oxford University biologist Richard Dawkins puts it, "a very special kind of explanation."

We would stagger in disbelief if someone told us that, in the fullness of time and inevitable course of things, singular objects such as artificial flies will as a matter of chance happen to land on one's windowsill, but there is no point in getting worked up about it; the pattern of dust motes on the sill is equally singular. Our reason justifiably rebels at the suggestion. "Universal experience," said William Paley in the late eighteenth century, "is against it." Artificial flies do not come into existence as a matter of chance. Yet what is true for the artificial fly holds *a fortiori* for ordinary flies. In neither case will we be content to ascribe the origin of these objects to the workings of chance.

This intuition—that even the simplest living things are "strange objects," in Monod's apt phrase—was something Charles Darwin imbibed from Paley and retained throughout the changes of outlook that marked his scientific life. The specified complexity of organisms, "that perfection of structure and coadaptation," Darwin wrote, "which most justly excites our admiration," could not be denied.[5] As Harvard evolutionary biologist Richard Lewontin observes, "Darwin realized that if a naturalistic theory of evolution was to be successful, it would have to explain the apparent perfection of organisms and not simply their variation."[6]

And not just any sort of explanation would do. Early in *Origin of Species*, after recounting the complex structures of the woodpecker and the mistletoe, Darwin gently mocks and then draws a sober epistemological moral from the failings of his contemporary and fellow evolutionist Robert Chambers:

> The author of the "Vestiges of Creation" [i.e., Chambers] would, I presume, say that, after a certain unknown number of generations, some bird had given birth to a woodpecker, and some plant to the mistletoe, and that these had been produced perfect as we now see them: but this assumption seems to me to be no explanation, for it leaves the case of the coadaptations of organic beings to each other and to their physical conditions of life, untouched and unexplained.[7]

Proper explanations appeal rather to *vera causa,* that is, to causes known to exist "independently of the facts they are supposed to explain."[8] Darwin's guiding uniformitarianism, imbibed from an even greater influence than Paley, Charles Lyell, would never allow the sudden birth of a fully formed woodpecker from a species of nonwoodpeckers. How could such an explanation be anything other than a miracle?

According to Darwin, a transformation broken into countless "infinitesimally small differences," however, or distributed across "many slight variations, accumulated during an almost infinite number of generations," might fall within the realm of possibility.[9] What one wants is a mechanism that will accumulate the useful or advantageous variations thrown up by nature and assemble them over time into novel structures or behaviors. Then, perhaps, one could pass between the horns of the dilemma "chance or design," by showing that natural processes can take the place of a designer. There would thus be no need to invoke an intelligence that originally contrived the specified complexity of living things.

Darwin found his transforming and designing mechanism in the principle of natural selection. As his notebooks make plain, Darwin came to natural selection after trying and rejecting several other mechanisms of evolutionary change. Drawing on an analogy with artificial selection, Darwin argued that natural forces could modify

131

permanently the characteristics of a species over time. We see several dozen sheep milling in a pen. A handful of males have fleece notice-ably thicker than the others. We remove these animals from the pen, and allow them—and no others—to breed with the estrous ewes. After the lambs are born, we look again for those with the thickest fleece and breed them. In several generations of this selective breed-ing the characteristics of the flock as a whole will have changed.

A string of unusually cold winters might cull a herd of sheep as effectively as any farmer, however, leaving only those animals with fleece thick enough to enable them to survive and reproduce. Indeed, nature's selective power might act on any variation, to the end of perfecting any structure or function far beyond what human power could accomplish. As Darwin argued in *Origin,*

> [A]ny variation, however slight and from whatever cause proceeding, if it be in any degree profitable to an individual of any species . . . will tend to the preservation of that individual, and will generally be inherited by its offspring. The offspring, also, will thus have a better chance of surviving, for, of the many individuals of any species which are periodically born, but a small number can survive. I have called this principle, by which each slight variation, if useful, is preserved, by the term of Natural Selection, in order to mark its relation to man's power of selection.[10]

The simplicity and commonsense appeal of this principle make it seem self-evidently the case. What could be more straightforward?

But Will It Fly?

But suppose we wanted to know whether this principle of nat-ural selection, at least as Darwin formulates it, were true. We should need to begin by identifying within a species the "profitable" or "useful" variations, which, by conferring an advantage on their pos-sessors, enable those organisms to succeed in the struggle for exis-tence, and to leave offspring likewise enabled. So how do we know which organisms carry the useful—i.e., the evolutionarily signifi-cant—variations?

Presumably these are not to be identified only after the fact by tallying their actual survival and reproduction, for then the phe-nomenon of "usefulness" would collapse into the phenomenon of "survival," i.e., reproduction. This would render Darwin's hypothesized cause, usefulness, analytically indistinguishable from its effect, survival and reproduction. If Darwin's theory turned out to be the claim that the useful variants are those that survive, the principle of natural selection could, unhappily, not fail to be true. That is, some living things do indeed survive and leave more offspring than others. But if this differential survival and reproduction is credited to the presence of "useful variations"—meaning only whatever "tends to the preservation of an individual"—any hope of explanation slips through our fingers. As the British biologist R. I. M. Dunbar notes, evolutionary explanations would then "be reduced to mere descriptions of observed fact. Statements that appear to offer explanations for the evolution of particular characters turn out on closer analysis to be no more than restatements in definitionally equivalent form of the facts that they purport to explain."[11]

Consider an example. Suppose we are observing a population of thirty geese of reproductive age, forming fifteen breeding pairs. We want to predict which of these pairs will leave the greatest number of offspring in the next generation. We want to do so, moreover, by reference to some heritable trait that the geese possess, because that trait is "useful" or "profitable," enabling some geese to survive better and leave more offspring than others. The trait, in short, confers a greater "evolutionary fitness" on its possessors.

Suppose we hypothesize that in our population of geese, body weight, an independently measurable trait, is the most important determinant of future reproductive success. Geese with greater body weights should produce more goslings, thus making them the fittest in the population.

Over several generations, however, we observe no consistent relation between body weight of parents and number of their goslings. Frustrated, we then hypothesize that, whatever their body weights, those geese producing the most offspring are the fittest. We then wait to see which geese, in fact, produce the most offspring.

But now cause and effect have collapsed into each other. As the philosopher of science Ronald Brady notes, when we observe that some members of a population leave more offspring than others, "what we want to know is *why*."[12] Such knowledge would have predictive value. To be told, however, that the *reason* they leave more offspring is *because* they leave more offspring is not instructive, nor is the circle broken by calling the most prolific members of the population "the fittest." As Brady elaborates,

133

Observation of natural populations reveals that some individuals leave more offspring than others. If we care to assign a *name* to those prolific individuals, that is our own convention. But if we now go further and say that the name explains *why* they leave more offspring, we have forgotten the factual basis of our identification of those individuals.[13]

What we need, it seems, is an empirical path independent of the actual facts of survival and reproduction for assessing what biologist J. G. Ollason calls "the quality of the phenotypes," that is, the actual organismal structures and behaviors—e.g., wings, eyes, nest-building, foraging patterns—directly "visible" to natural selection:

Underlying differences in fitness . . . must be differences in the phenotypes of the members of the lineages, the members of fitter lineages possessing phenotypes that are of better quality than the phenotypes of less fit lineages. If it were possible to define the quality of the phenotypes in some objective way the fitness problem would be completely solved, because then the process of natural selection would operate as Darwin proposed that it did: Because they are of a higher quality and quality is heritable, some individuals leave a disproportionately large number of offspring.[14]

What we want is something like an evolutionary "engineering analysis," as Harvard geneticist Lewontin terms it, to establish which phenotypes are of higher quality. Under such an analysis, we would find that some living things are relatively better engineered than others, that is, better able than their conspecifics to solve the problems posed to them by the environment. "An engineering analysis," argues Lewontin,

can determine which of two forms of zebra can run faster and so can more easily escape predators: that form will leave more offspring. . . . The concept of relative adaptation removes the apparent tautology in the theory of natural selection. Without it the theory of natural selection states that fitter individuals have more offspring and then defines the fitter as being those that leave more offspring; since some individuals will always have more offspring than others by sheer chance, nothing is explained. An analysis in which problems of design are posed and characters are understood as being design solutions breaks through this tautology by predicting in advance which individuals will be fitter.[15]

The escape from circularity will, however, be short-lived. To what end (in evolutionary terms) does the zebra run from the predator, or the sparrow forage, or any other organism ever do what it does? "The real 'problem' posed by the environment," argue the British

134

biologists Peter Saunders and M. W. Ho, "is purely and simply that of surviving and leaving offspring"—and a phenotypic structure or behavior "is important only insofar as it promotes this."[16] As far as evolution is concerned, a phenotype of the very highest quality that leaves no offspring has come (absolutely) to a dead end. Thus, any evolutionary engineering analysis must necessarily return to the actual facts of survival and reproduction. As Ronald Brady observes,

> The individual trait must be summed in the whole before we know how useful it actually is. Since the summing is beyond the knowledge of the investigator, he does not derive survival from his knowledge of engineering; he observes the fact of survival and then attempts to explain this by reference to design. How do we know that an animal is optimally designed for an environment? It survives in that environment. Thus, no matter how we explain good design after the fact, the criterion used for the detection of good design is always survival.[17]

What makes this problem so vexatious is the strong and widely shared conviction that some living things are really of "higher quality" or "fitter" than others. Surely we can get at this notion, a three-way relation between the organism, its environment, and other slightly different organisms, without sliding into a morass of circularity.

Here I suggest it is the conviction itself that bears examination. If buried within it lies the Darwinian premise that greater "evolutionary fitness" *explains* higher reproductive output, the slide into circularity is inevitable, for in retracing our steps we shall find that we can define evolutionary fitness only in terms of reproductive output. Even a cursory reading of Lewontin's writings on natural selection shows that he was well aware of this difficulty, although he stops short of Brady's conclusion, calling evolutionary engineering analysis "a tricky game." As Saunders and Ho observe, however, knowing how to score winners and losers in the evolutionary game isn't tricky at all. "The real criterion for choosing a solution [i.e., a successful design]," they note, "must be not how well it solves the sub-problem but how much it contributes to the central one, that of survival and reproduction."[18]

J. G. Ollason forcefully comes to grips with the difficulty. "The problem with evolutionary fitness," he argues, "is that there is no possibility, in principle, of establishing a mapping, in physical and chemical terms, from the phenotypic properties of the animal to its reproductive output, the main reason for this being that there is no way to define the quality of the phenotype."[19] If natural selection is formulated as Darwin understood it, the theory is trapped in an episte-

135

mological thicket from which it cannot be extracted. "Fitness"—namely, that quality by which some organisms supposedly outperform others in the evolutionary competition—is a deeply problematical notion. The problematic status of "fitness" as an operational construct was well described in 1989 by the University of Chicago evolutionary theorist Leigh Van Valen. "Yes, fitness is the central concept of evolutionary biology," writes Van Valen,

> but it is an elusive concept. Almost everyone who looks at it seriously comes out in a different place. There are literally dozens of genuinely different definitions, which I won't review here. At least two people have called fitness indefinable, a biological primitive. (A primitive is an undefined initial term in logic). I don't think that helps. Stearns (1976) once described it as "something everyone understands but no one can define precisely." Or is it that we can't define it because we don't fully understand it?[20]

Running in Circles

None of this should surprise any evolutionary biologist. The so-called tautology problem with natural selection, which we have been analyzing, has spawned a truly vast literature from biologists and philosophers, and many attempts to reformulate the principle. The University of Wisconsin philosopher of biology Elliott Sober quipped in 1984 that when "philosophers say they are writing a paper on 'the structure of evolutionary theory,' they mean that they are writing a paper on the tautology problem."[21] Within the past decade or so, however, evolutionary biologists are just as likely to abandon the task of reformulating natural selection, dispirited by continuing uncertainty over what logical form the principle should take, whether it is testable, or what it yields of explanatory value.

Consider logical form. While most biologists take pains to formulate natural selection precisely to avoid circularity, others argue that in its "deep" axioms, evolutionary theory is and, indeed, must be tautologous. Under the heading "Is circularity always a bad thing?" the Swiss biologists Stephen Stearns and Paul Schmid-Hempel write, for instance, that

> [i]n evolutionary theory, the deep axiom is that types with superior reproductive performance will be better represented in future generations. Because the statement is tautological, it is useful, for the tautology is a guar-

136

antee of logical consistency at the deepest level of the theory. The question here is really about the consistency of theories, where tautology is not only appropriate but precisely what one seeks to demonstrate through carrying out a proof.[22]

Or consider the question of testability. In response to the criticism that natural selection, while not formally tautologous, nevertheless makes no testable predictions, some have argued that the principle does yield predictions. The Canadian evolutionary biologists Bruce Naylor and Paul Handford, for instance, in a spirited defense of neo-Darwinism, allow that "although Darwinian theory is not designed to make predictions about the specific future form of organisms and events, it can, in principle, do so." They suggest a potential test case in the well-known phenomenon of resistance to the insecticide DDT:

> The evolution of DDT resistance has been observed in over 200 species of insects and other arthropods (Brown 1967). Predictions about the outcome of continued DDT use can be made: either all the lineages will become extinct, or some mechanism will arise by which the mortal effects of DDT are avoided.[23]

Reasoning of this sort—the lineages will become extinct or they won't become extinct—has led many evolutionary biologists (particularly those who completed their graduate training in the late 1960s or early 1970s) to regard the entire theory of natural selection with open cynicism. The paleontologist Donn Rosen of the American Museum of Natural History, for instance, urged that natural selection—whose "axiomatic nature has been virtually overlooked"—be sent off "to join the ether, phlogiston, and noxious vapors."[24] Joel Cracraft of the University of Illinois is equally dismissive. "As far as I can see," he wrote in 1981, "statements of the type that 'phenotype x is an adaptation, evolved via the agency of natural selection,' are thoroughly untestable. The necessary data needed to refute such an assertion cannot be gathered, and we are more or less forced to accept it as an article of faith rather than as a scientific statement."[25]

Doubtless the most significant criticism to emerge in the skeptical reevaluation of natural selection, however, concerns its putative creative role in the history of life. Defenders of neo-Darwinism have sometimes objected that critics heedlessly run natural selection and evolutionary theory together, so that difficulties with conceiving the empirical content of the former tell against the soundness of the latter. Sober, for instance, in response to the tautology problem, worries that "the problem has frequently been inflated, so that the entire empirical status of evolutionary theory is seen to stand or fall on the status of a single proposition. But why is it important whether this or that proposition is empirical?"[26]

The tautology problem is important, it should be plain, because of what has historically been claimed for natural selection. Here the intellectual sight of many evolutionists often seems to be clouded: they cannot seem to remember exactly what has been credited to the principle of natural selection, or to some other cause, or to evolution in general. Therefore I provide the following, beginning with Darwin himself: "I am fully convinced that species are not immutable. . . . Furthermore, I am convinced that Natural Selection has been the main but not exclusive means of modification."[27]

As expressed by the distinguished SUNY-Stony Brook theorist George Williams: "Natural selection . . . [is] the only acceptable theory of the genesis of adaptation."[28]

Or by leading neo-Darwinians Dobzhansky, Ayala, Stebbins, and Valentine: "According to the theory of evolution . . . natural selection is the process responsible for the adaptations of organisms, and also the main process by which evolutionary change comes about."[29]

Or by the English evolutionary theoretician John Maynard Smith:

> The fact of evolution was not generally accepted until a theory had been put forward to suggest how evolution had occurred, and in particular how organisms could become adapted to their environment; in the absence of such a theory, adaptation suggested design, and so implied a creator. It was this need which Darwin's theory of natural selection satisfied. He was able to show that adaptation to the environment was a necessary consequence of processes known to be going on in nature.[30]

Or by his countryman, Richard Dawkins: "Adaptation cannot be produced by random drift, or by any other realistic evolutionary force that we know of save natural selection."[31]

Or, lastly, by the dean of neo-Darwinians, zoologist Ernst Mayr of Harvard: "The real core of Darwinism . . . is the theory of natural selection. This theory is so important for the Darwinian because it permits the explanation of adaptation, the 'design' of the natural theologian, by natural means, instead of by divine intervention."[32]

I have belabored this point to emphasize that, if natural selection should fail, the standing of the neo-Darwinian story cannot escape

138

the damage. Natural selection is not a peripheral aspect of neo-Darwinism. The principle lies at the very center of the theory.

Redesigning Natural Selection

In light of this, one might admit to some surprise on reading that the principle of natural selection "is not an explanation for adaptation," "addresses the problem of the *spread* of new variants or new adaptations, not their *origin*," and causes "only frequency changes in populations."[33] These statements are taken from a 1986 treatise, *Natural Selection in the Wild*, by the University of California geneticist John Endler. How could Endler, a neo-Darwinian, hold such a radically different opinion of selection from Darwin, Mayr, Maynard Smith, and the others?

Endler undertook his treatise on selection, he writes, to show that "it is neither a tautology nor a metaphysical exercise."[34] To avoid falling into familiar difficulties with the concept of fitness (which had led to charges of tautology), Endler painstakingly reformulated the principle of natural selection. In so doing, however, he fundamentally shifted the content of the principle.

Recall that Darwin said that the "useful" variants would be those tending to favor the survival (and reproduction) of their possessors. This quality of "usefulness" disappears in Endler's formulation for precisely the reasons sketched above—it is impossible to measure independently of actual survival and reproduction. In its place, Endler erects the following syllogism:

If, within a species or population, the individuals

a. vary in some attribute or trait q (physiological, morphological, or behavioral)—the condition of **variation**;
b. leave different numbers of offspring in consistent relation to the presence or absence of trait q—the condition of **selection differences**;
c. transmit the trait q faithfully between parents and offspring—the condition of **heredity**;
d. *then* the frequency of trait q will differ predictably between the population of all parents and the population of all offspring.[35]

Endler stresses that conditions a, b, and c are jointly necessary and sufficient for natural selection d, to occur. (The process of genetic drift, Endler also points out, differs from natural selection in that condition b is absent, by definition, under random sampling.) Now, while this formulation may seem superficially to resemble Darwin's own, it actually differs profoundly.

First, as noted, Endler's formulation successfully avoids the problem of circularity—more precisely, the collapse of "usefulness" (i.e., fitness) into "survival and reproduction"—by rigorously excluding any assessment of what Ollason calls "the quality of the phenotype," i.e., how exactly it is that one variant competes better than another by being "better adapted." In so doing, however, Endler strips natural selection entirely of the great intuitive force Darwin originally gave it. The bird was given its wing by selection because wings are useful in the struggle for existence. Bird's wings are adaptations, and adaptations are produced by natural selection. And so on.

Stop, stop! Endler shouts (in effect)—we've been down that road already, and it leads inevitably to a morass. "To say that a new adaptation necessarily arose through natural selection," he writes, "is an incomplete description, a tautology, and a misrepresentation of natural selection, adaptation, and evolution."[36] Properly to infer the action of selection, Endler stresses, we must have knowledge of conditions (a), (b), and (c), *and these exhaust the content of the principle.*

Redesigning Design

This point—that natural selection is simply a directional shift in the trait frequencies of species—was not what Darwin, or modern neo-Darwinians, envisioned for the principle. And this leads to the second, and more important, difference between Endler's sound formulation of selection and the flawed formulations. In Darwin's view, selection was the primary crafter of the amazing design of organisms. All that quasi-agency disappears in Endler's formulation. Our attention is focused instead on the requirement that we give evidence for (a) variation, (b) a consistent relation between variation and offspring, and (c) heritability. *Absent that evidence, natural selection has no other content.*

This can be illustrated with a final example. Dawkins has argued that "the theory of natural selection provides a mechanistic, causal

140

account of how living things came to look as if they had been designed for a purpose."[37] Thus, it seems reasonable to suppose that, if we select a particular structure—such as that trusty standby from natural theology, the vertebrate eye—Endler's formulation above should point us in the direction of a "mechanistic, causal account" for the origin of that structure.

Suppose, then, that we wished to explain how the vertebrate eye came to be: just the sort of explanatory role natural selection putatively handles well. This task was taken up in some detail by Carl Gans and R. Glenn Northcutt, in 1983, in their theory on the origin of the anatomical structures peculiar to vertebrates. The fossil evidence is not helpful—"as the first recognizable agnathans [vertebrates]," write Gans and Northcutt, "already had all major sense organs, brain divisions, and cranial nerves"[38]—but using existing species, Gans and Northcutt developed the following scenario, which I have separated by numerals into its major hypotheses. Now, the vertebrate eye is found in the vertebrate head, wired up to the vertebrate brain; so, sensibly, Gans and Northcutt first give an account of the origin of the vertebrate head:

1. The primary shift to a vertebrate condition apparently involved a modification of the filter-feeding mechanism to a mechanism that permitted use of selective predation and made larger prey items accessible.
2. To the extent that larger prey was available in the environment, its capture would not only have increased the range of nutrients, but established advantages for increased predator size and consequently for increased metabolic output; an improvement of the mechanism for gas exchange would then have acquired a substantial advantage.
3. This improvement was accomplished by muscularization of the hypomere into "branchiomeric" muscle and by the capacity for muscular deformation of the pharynx.
4. Also, there was replacement of collagenous pharyngeal bars by more elastic cartilaginous ones that permitted elastic recoil, using the energy stored during muscular pharyngeal deformation.
5. In parallel, there was subdivision of the circulatory system into capillary beds beneath the gill epithelia, muscularization of the aortic arches, the development of a central heart and circulating erythrocytes.
6. Also, the wall of the gut became muscularized and this increased the capacity to deal with larger prey items in a larger lumen utilizing extracellular digestion.
7. All of these changes involved the development of new sensory, integrative, and motor controls, which apparently were centralized by expanding the neural tube to form the spinal cord and hindbrain.

141

8. At this stage the paired and external special sense organs developed, as well as a central integrative capacity for utilizing the increased information they provided.[39]

What is the explanatory role of natural selection in this rather complex scenario? In the second and third hypotheses, there is some passing mention of "selective advantage"—but these advantages, like the variations and transformations on which they depend, are entirely hypothetical. Here, natural selection is nothing more than a deductive formula through which a great number of hypothetical evolutionary variables are being run.

That is, if it were the case that a species of "hypothetical proto-vertebrate" (Gans and Northcutt's term):

- varied sufficiently in its musculature (hypothesis 3) and collagenous tissues (hypothesis 4) to evolve an elastic pharynx *de novo;*
- varied sufficiently in its circularity system to evolve capillary beds, muscular aortic arches, a central heart, and circulating erythrocytes (hypothesis 5) *de novo;*
- varied sufficiently in its digestive system to evolve a muscular gut with a larger lumen or cavity (hypothesis 6) *de novo;*
- varied sufficiently in its nervous system to evolve new sensory, integrative, and motor controls by evolving a spinal cord and hind brain (hypothesis 7) *de novo;*

then the vertebrate eye might have evolved.

The "mechanistic causal account" promised by Dawkins is in fact a long chain of suppositions, to whose truth or falsity the innocuous syllogism of natural selection is wholly irrelevant. "Nothing can be effected" by natural selection, Darwin wrote, "unless favorable variations occur."[40] Whether such variations have, or could have, occurred are factual questions to which selection is helpless to speak. Indeed, as Endler argues, "The fundamental mechanisms of evolution are the molecular mechanisms leading to new genetic variants, the expression of those variants through the genetic and developmental systems, and constraints to the appearance and function of those variants."[41] This knowledge must be gathered *before* natural selection can be invoked; and it is here, in understanding the precise mechanisms of variation via development, that evolutionary biology has been least successful.

A Placeholder for the Unknown

We might conceive of the principle of natural selection as a simple deductive machine whose fuel is the observed evidence of heritable variation. Without that fuel, the machine sits idle. Swedish evolutionary biologist Soren Løvtrup noted in 1979 that "without variation, no selection: without selection, no evolution. This assertion is based on logic of the simplest kind, and it should be noted that the common implication of selection pressure as an evolutionary agent becomes void of sense unless the availability of the proper mutations is assumed."[42] As the philosophers of biology Michael Bradie and Mark Gromko argue,

> The principle of natural selection explains the evolutionary history of a given species or form only to the extent that specific characteristics of the organism in question and specific factors in the environment are determined. The principle of natural selection is, thus, to be interpreted as an existential claim to the effect that such characteristics and factors exist. Until the specific characteristics and factors, which are relevant for a given situation, are isolated, the principle of natural selection provides, at best, only an explanatory sketch of the evolutionary process in question.[43]

"Natural selection *per se*," writes University of Minnesota philosopher of science Arthur Caplan, "explains nothing."[44] This insight is put at risk, however, by his amplifying statements:

> Rather, natural selection is a useful label for referring to an extraordinarily complex array of causal interactions occurring at the level of genes, genotypes, phenotypes, and environments. It is the laws and generalizations of genetics, development, ecology, and demography, which ultimately are invoked by biologists to explain change and descent in the history of life. Natural selection is simply a covering term or place-holder for describing the various processes involved in producing evolutionary change, or the products of such change.[45]

But, as many evolutionary biologists have come to realize, natural selection does not serve as a placeholder where the real evidence has stepped away for a moment. Rather, it is precisely because the evidence (of large-scale heritable variation) is *wanting* that natural selection indeed explains nothing. Thus, those biologists who argue, like the Austrian evolutionary theorist Gerhard Müller, that "the origin of new morphological characters is still unexplained by the current synthetic theory,"[46] have not neglected to read Darwin or Dawkins.

Rather, they have gone there, weighed natural selection for what it truly is—a simple deductive formula—and come away empty-handed.

Darwin himself, for his many insights into the nature of living things, always assumed much that his followers would not be allowed to assume. It fell to the neo-Darwinians to explain the origin of the manifold designs of living things. In a letter to the geologist Charles Lyell, one of his closest friends, Darwin wrote:

> I have reflected a good deal on what you say on the necessity of continued intervention of creative power. I cannot see this necessity; and its admission, I think, would make the theory of Natural Selection valueless. Grant a simple Archetypal creature, like the Mud-fish or Lepidosiren, with five senses and some vestige of mind, and I believe natural selection will account for the production of every vertebrate animal.[47]

A mudfish is not, however, a simple creature; and postulating the existence of the very thing one wants to explain—e.g., "five senses"—gives away the game. Either natural selection can do the work of building organisms from scratch or it cannot. I submit that it cannot, that evolutionary biologists are coming to understand this, and therefore that the problem of design—the problem of the origin of ordinary flies, if you will—remains open and unsolved.

11

THE CAMBRIAN EXPLOSION

The Fossil Record and Intelligent Design

ROBERT F. DEHAAN
AND JOHN L. WIESTER

Robert F. DeHaan, Ph.D. (human development, University of Chicago), is retired; he taught developmental psychology at the University of Chicago and Hope College.

John L. Wiester (B.A., geology, Stanford University) is chairman of the Science Education Commission of the American Scientific Affiliation, an association of Christians in the sciences, and has taught geology at Biola University for the past five years. He is the author of The Genesis Connection[1] *and coauthor of* Teaching Science in a Climate of Controversy,[2] *as well as* What's Darwin Got to Do with It?[3]

Charles Darwin's *Origin of Species* drastically changed the way Western culture views design in nature. Before the publication of that great work, it was commonly assumed that the wondrous complexity of living things required a designing intelligence. That is no longer the case. Darwinists claim that the obvious design of biolog-

145

ical organisms is not real but only apparent. Francisco Ayala, in 1994 the president of the prestigious American Association for the Advancement of Science, put it this way:

> The functional design of organisms and their features would therefore seem to argue for the existence of a designer. It was Darwin's greatest accomplishment to show that the directive organization of living beings can be explained as the result of a natural process, natural selection, without any need to resort to a Creator or other external agent. . . .Darwin's theory encountered opposition in religious circles, not so much because he proposed the evolutionary origin of living things (which had been proposed many times before, even by Christian theologians), but because his mechanism, natural selection, excluded God as the explanation accounting for the obvious design of organisms.[4]

That is a monumental claim. Ayala is not simply talking about natural selection producing minor variations, like changes in the shape of finch beaks as an adaptation to the kind of food that finches eat, or like the reversible camouflage coloration of guppies to protect them from predators. He means that natural selection is sufficient to account for *all* aspects of living things, from the largest innovations to the smallest variations. He claims that natural selection has banished intelligent design from the halls of science and dumped it in the dustbin of history.

Since natural selection operates on random mutations in genetic material, within Darwin's theory *chance* lies at the very foundation of life. Chance thus supplants purpose, which had been the pivotal feature of intelligent design. In the pre-Darwinian view, life was planned and purposeful. In the Darwinian view, life arose and evolved solely by what Ayala calls "the creative duet of chance and necessity," without purpose or a "preconceived design."

Julian Huxley went even further, pushing Ayala's biological evolution to the cosmic scale. At the University of Chicago's centennial celebration of the publication of Darwin's *Origin of Species* in 1959, Huxley claimed that the entire cosmos came into being and continues to exist under the direction of natural selection operating on chance processes. And if the entire cosmos, said Huxley, why not all of human history and culture as well? The sweep is complete. Chance and necessity rule everything. Intelligent design thus becomes sidelined in any discussion of biology, cosmology, history, and culture. According to Huxley, design is utterly passé.

146

The Testimony of Fossils

Are Ayala and Huxley right? This essay argues that they are not. The purpose of this essay is two-fold: first, to show how utterly mistaken it is to claim that natural selection provides the sole organizing principle in the history of life and especially in the formation of major innovations; and second, to show the necessity of intelligent design in any plausible account of major innovations.

The key question, therefore, is the following: Can natural selection explain the origin of *major innovations* in living things—like the initiatory body plans of large animal groups, the emergence of new organs, shapes, and functions in plants and animals, and the hierarchical organization and discontinuities among large groups of animals? And can it do so without invoking design?

The fossil record provides the scientific evidence needed to answer this question. Fossils are the mineralized remains, impressions, and traces of animals and plants that lived long ago that were preserved and deposited in the layers of rock that form the earth's crust. Dinosaurs are the current darlings of the fossil record. Even so, there is much more to fossils than dinosaurs.

To determine whether natural selection really does explain "the directive organization of living beings," as Ayala asserts, let us begin with predictions by Darwin and his followers about what one should expect to find in the fossil record if Darwin's theory were true. Darwinism's explanatory mechanism of natural selection operating on random variations predicts the following:

1. Major lineages or groupings of animals should start with varieties and species that would be modified by natural selection, and after perhaps many thousands of generations, would form ever more divergent groups, until eventually unique groups at the highest organizational levels would be formed. That is, *diversity* (varieties and species) should precede and result in *disparity* (unique, nonoverlapping groups of animals).
2. This process of diversification resulting in disparity should be accompanied by numerous intermediate species, transitional forms, and failed experiments, the by-products of natural selection.
3. New major groupings of animals (called phyla) should emerge from time to time as one moves up the geologic time scale, and they should increasingly diverge from each other.

147

To verify these Darwinian predictions, it is necessary to go back about 530 million years in the fossil record to an event so important that it is called the watershed in the history of life, second in significance only to the origin of life itself—the Cambrian explosion, called by some "Biology's Big Bang." The Cambrian explosion refers to the geologically sudden appearance of multicellular animals during the geologic period called the Cambrian.

The entire period of the Cambrian explosion spans no more than 5 to 10 million years. In this small window of time fifty or so unique animal body plans appear, most of which are not found earlier in the fossil record and many of which persist to this day. That's not to say nothing was happening in the time leading up to the Cambrian explosion. But whatever was happening in the Precambrian eon wasn't drawn together, augmented, and organized until the Cambrian explosion. It took the Cambrian explosion to assemble fifty radically different animal types, each one the progenitor of a major theme of life's history.

This event occurred suddenly—in a geologic eye blink. It was never to be repeated. Most significantly, it produced an extraordinary breadth of basic animal designs. From these Cambrian animals issued nearly all the major animal groups (or "phyla" as they are known technically) that ever existed on earth.

Each of the fifty Cambrian animal types can be identified by its unique body plan. To see what's at stake here, consider an automobile. Automobiles have a basic body plan—wheels, axles, a chassis, body, power supply, and stopping and steering mechanisms. Each new model adds modifications, but the basic body plan is never abandoned. Next consider an airplane. Its basic architecture is as fixed as that of an automobile, yet it is radically different from an automobile. As a third example think of a sailing vessel. All of these have basic designs that are unique. True, they may borrow specific components from each other, but the basic design remains constant regardless of subsequent developments.

Evidence Versus Theory

So it is with the basic biological body plans of the Cambrian period. Thirty-seven of those body plans have survived to this day. These have been passed on from generation to generation over hundreds of millions of years. They have been elaborated with additional features but never basically altered. For example, there was an animal named Yunnanozoon (found in the Chengjiang fauna in the Province of Yunnan in southern China) that appeared in the Cambrian explosion. Its body plan consisted of a notochord (which in later generations became the backbone that encases the spinal cord) and striped muscles. These two characteristics show up in all the descendants of Yunnanozoon and are the main identifying characteristics of the phylum Chordata to which the subphylum Vertebrata belongs. New structures were added along the way—fins, subsequently four appendages, and still later fur and mammary glands. But even after hundreds of millions of years, none of the progeny ever lost the vertebral column, the hallmark of the vertebrates.

Do the Cambrian animals and their progeny bear out the three predictions given above—the predictions of what the fossil record should look like if Darwinism were true? Are these predictions borne out by the evidence of the fossil record and specifically by the evi-

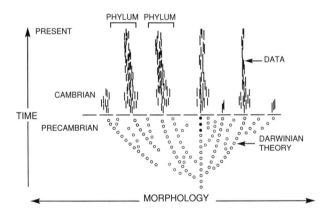

Figure 1. Darwinian Theory vs. the Fossil Evidence.

dence of the Cambrian explosion? Let us start with the primary division among animals—the phyla. Do they originate in a long spreading radiation of species as predicted by Darwinian theory?

Consider Figure 1. It depicts Darwinian theory in relation to the fossil evidence. Along the bottom axis is the range of morphology. Along the left hand margin is the timeline from approximately 1,200

149

million years ago in the Precambrian period up into the Cambrian period. A major geological dividing line at 545 million years ago separates the Precambrian eon from the Cambrian period. The major biological dividing line occurred 530 million years ago during the Cambrian period and thus is called the Cambrian explosion.

Darwinian theory predicts that from a single progenitor would bubble up, as it were, numerous branching lineages as shown by the circles in the Precambrian eon. These would then be the precursors of the Cambrian animals. By contrast, the data show that virtually all phyla originated in the Cambrian explosion, with a probable fuse leading to the explosion extending back into the Precambrian eon. Note that the phyla have narrow bases and few, if any, connections to hypothetical Darwinian precursors.

The Cambrian explosion marks what Niles Eldredge called, "the relatively abrupt appearance of complex animal life that marks the beginning of a rich and dense fossil record."[5] While he claims that it is "a fascinating example of the phenomenal speed at which evolution can work," it is clearly contrary to the Darwinian prediction that changes wrought by natural selection would be slow and gradual. In sum, three characteristics mark the Cambrian explosion from all other events in the history of life: Its speed (5–10 million years); its breadth (50 disparate animals that were the progenitors of the large groupings of animals called phyla); and its finality (only one additional phylum formed after this time).

Moreover, each phylum is self-bounded. Indeed, there are no transitional forms between them, as predicted by Darwinian theory. The Darwinian mechanism of selection and variation does provide a plausible explanation for minor variations among species, such as the varieties of shapes in finches' beaks. But this mechanism plays no discernible part in the formation of major innovations.

Does the number of phyla increase over time, as Darwinian theory predicts? Figure 2, titled "Origin of the Phyla," provides the answer. According to the top graph, labeled "Darwinian Predictions," the number of phyla should increase over time. Driven by the ubiquitous process of natural selection, animal evolution should exhibit no end of novelties. The bottom graph labeled "The Fossil Evidence," however, shows that at 530 million years ago the number of phyla increased sharply to fifty in less than 10 million years, and thereafter dropped to thirty-seven at the present time. At the

Darwinian Predictions

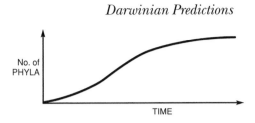

The Fossil Evidence

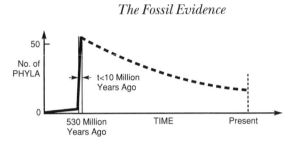

Figure 2. Origin of the Phyla.

very place where the scientific evidence of the fossil record should support the Darwinian prediction most clearly, namely, at the origin of the highest category of animals, there the Darwinian theory breaks down completely.

The Discontinuity and Hierarchy of Life

What has happened in the history of life on earth since the Cambrian explosion? The distinguished evolutionist Theodosius Dobzhansky summarized two outstanding characteristics of the diversity in the living world—its *discontinuity* and its *hierarchical organization*. As for discontinuity, life is not a hodgepodge of plants and animals that appeared randomly in the history of life, as one might expect if natural selection were the sole organizing process. Rather, the fossil record reveals that plants and animals have been organized from the very outset into distinct groups, some of which have existed

151

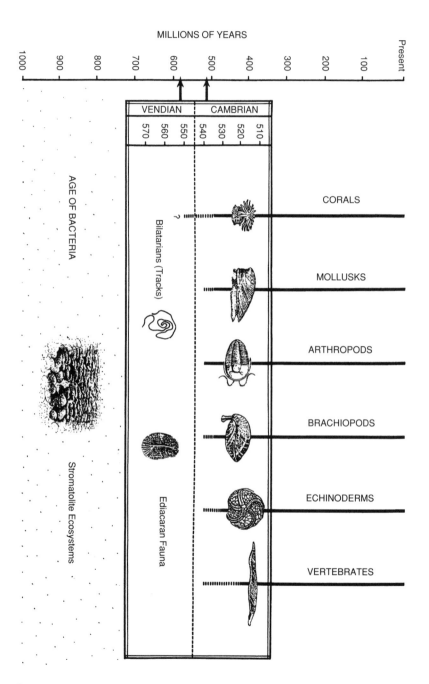

Figure 3.

for millions of years. These groups do not overlap, which is to say they are discontinuous.

One large group of animals, as already discussed, possesses a backbone, or a vertebral column encasing the spinal cord. The group includes all animals from fish to amphibians to reptiles to mammals, and thus includes human beings. That group quite appropriately is called vertebrates. A different major division of animals does not have a backbone and internal skeleton. The animals in that division have an external skeleton or hard outer covering. They are called invertebrates, and include insects, crabs, lobsters, and other animals with an external skeleton. There are thirty-seven phyla, which, besides those already listed, include worms, jellyfish, corals, and mollusks. They all have completely different body plans.

The discontinuities among major design themes in the fossil record mirrors the discontinuities among major design themes in technology (cf. the differences between automobiles, airplanes, and computers). The major themes may borrow elements from each other, but they never lose their unique identity.

The second major characteristic of animals and plants, according to Dobzhansky, is that each major group developed over geological time hierarchically—like a pyramid. However, unlike pyramids, which are built from the bottom up, phyla are constructed from the top down. As shown in Figure 3, each phylum started with a single progenitor at its head, which had certain characteristics that it passed on to all its progeny. At each lower level in the hierarchy, the populations added new structural innovations and became more numerous and diverse. Yet none of the generations of animals ever lost the defining characteristics of the phylum, inherited from the ancestral Cambrian animal.

The hierarchical organization of life makes it possible for scientists to classify animals and plants in an orderly fashion. The hierarchical pattern among living things, in which a major design theme first originates and then is followed by variations on that theme, mirrors the hierarchical pattern among technological devices, in which, likewise, a major design theme first originates and then is followed by variations on that theme (cf. the different styles of automobiles).

Does the fossil record exhibit a hierarchical pattern? There is abundant scientific evidence that animal groups are hierarchically organized from the top down. This top-down pattern is repeated with different groups of animals in the fossil record. The fossil evidence therefore indicates that natural selection was not a major player at the upper (earliest) levels of this top-down process.

Since these hierarchies have their origin in the unique body plans of the Cambrian animals, and since there is no substantive evidence that the Darwinian mechanism or any other naturalistic processes had a hand in the formation of body plans, we are under no obligation to conclude that brute undirected naturalistic processes underlie the formation of hierarchies and discontinuities in the world of living things. Chance and necessity do not appear to be the decisive causal factors in the emergence and development of life.

The Case for Cambrian Design

But can the discontinuities and hierarchies in the living world be considered evidence for intelligent design? Does the scientific evidence from the fossil record warrant a design inference? We submit that it does. Intelligent design is purposeful and forward acting, and capable of providing structure and organization to biological processes. These features of intelligent design apply readily to the Cambrian explosion:

1. The Cambrian explosion itself is an expression of design. It brought order out of the melange of biological events of the Precambrian period by assembling animals into fifty lines of development, thirty-seven of which have continued to this day.
2. Each phyletic hierarchy is defined by the body plan of the Cambrian animal that was its progenitor, and maintained the integrity of the basic design throughout hundreds of millions of years. The discontinuities have also been maintained throughout the history of life since the Cambrian explosion.
3. The body plans could just as legitimately be called body *designs*, indicating that their function was purposeful and ensuring that the hierarchical organization and the discontinuity of phyla became permanent features in the organic world.
4. The phyletic lineages that originated in the Cambrian explosion are arranged hierarchically and have all followed a consistent track of development over the geological ages.

154

Does natural selection explain, as Ayala asserts, "the directive organization of living beings"? We have presented data that cast serious doubts on Ayala's assertion. The same data provide compelling evidence of design and show the necessity of incorporating intelligent design into any theory that attempts to account for major biological innovations, like those in the Cambrian explosion.

Even so, the myth continues that Darwinian theory is not just true, but also unassailable and impregnable. History, however, teaches that seemingly invincible theories are nothing of the sort. In the middle of the nineteenth century a theory was proposed to answer the following question in geology: How did major mountain ranges originate? The geosynclinal theory hypothesized that huge trough-like depressions, known as geosynclines, became filled with sediment and subsided until they gradually became unstable; then, through heat from the interior of the earth, the sediments became crushed, folded, and elevated into a mountain chain. This theory was so well established that as late as the 1960 edition of Clark and Stearn's *Geological Evolution of North America*,[6] the status of the geosynclinal theory was compared favorably with Darwin's theory of natural selection. The book asserted:

> The geosynclinal theory is one of the great unifying principles in geology. In many ways its role in geology is similar to that of the theory of evolution, which serves to integrate the many branches of the biological sciences. . . . Just as the doctrine of evolution is universally accepted among biologists, so also the geosynclinal origin of the major mountain systems is an established principle in geology.[7]

Yet, within the short span of only ten years following the publication of that geology book, the geosynclinal theory was replaced by the theory of plate tectonics, which combined the hypotheses of continental drift and sea floor spreading to explain mountain formation. The new theory held that continents, after drifting inexorably toward each other, crashed, resulting in the upraised mountain ranges found on the continents. The Himalayan Mountains, for instance, are rising because the Indian subcontinent is slowly crashing into the Asian continent.

The theory of Darwinian evolution is no more impregnable than the old geosynclinal theory of mountain formation. The theory of intelligent design, with its new perspective on design in biology, is destined to replace Darwin's mechanism of natural selection. The implications of this paradigm shift are staggering. No longer will natural selection be assigned a creative role in the formation of major innovations in the history of life and of the cosmos. Instead, intelligent design will become discernible and pervasive throughout the natural order. This will radically alter Western culture's outlook not only on science, but also on the very nature of reality.

12

THE "JUST SO" UNIVERSE

The Fine-Tuning of Constants and Conditions in the Cosmos

WALTER L. BRADLEY

Walter Bradley, Ph.D. (materials science and engineering, University of Texas at Austin), is professor of mechanical engineering and director of the Polymer Technology Center at Texas A&M University. He is the author of over one hundred publications in materials science and coauthor with Charles Thaxton of The Mystery of Life's Origin, *a seminal work on the chemical basis of life. He is a fellow of the American Society for Materials and the American Scientific Affiliation.*

*W*hat does it mean on a human scale for an engineer to design a product? What would it mean on a grand scale to say that the universe is the product of an intelligent designer? And what evidence could support such a claim? What features of the universe suggest that a "home" has been carefully crafted for our benefit?

William Paley, in his classic *Natural Theology*,[1] provided evidence from both the physical sciences and the biological sciences for a designed universe, but the strength of his argument for design was limited by the scientific understanding of his time and was subsequently called into question by Darwin's theory of evolution. However, discoveries in astronomy and cosmology in the last half of the twentieth century have provided extremely compelling evidence for a designed universe. Before we look at the evidence from cosmology indicating that our universe is indeed designed as a habitat for life in general and humans in particular, we need to clarify what we mean by design.

How Does an Engineer Design Something?

To understand what engineers do when they design products for consumers, consider first a simple example of how we guide physical events to accomplish a purpose. Suppose that I wanted to throw a water balloon from the Leaning Tower of Pisa in Italy to the plaza below, hitting a friend who is walking on the plaza (and missing other tourists). Using the equations Newton discovered for motion and for gravitational attraction, I could describe the descent of the water balloon to the plaza below with the following simple algebraic relationship:

$$H(t) = h_0 - (G\,m\,/\,r^2\,)\,t^2\,/2 - v_0\,t$$

Here G is a universal constant that gives the strength of the gravitational force of attraction, m and r are the mass of the Earth and the radius of the Earth, and h_0 and v_0 are the height in the tower from which I shall throw the balloon and the vertical velocity with which I throw the balloon.

With these constants and initial conditions defined, I can then calculate the height of the water balloon. $H(t)$ gives the calculated height of the water balloon as a function of time t from when I threw it. This equation may be used to guarantee that my balloon arrives at the plaza at just the right time to hit my friend. All I need to do is to determine at what time my strolling friend will be just below me in the plaza, and then I can use the equation to determine the initial velocity with which I need to throw the balloon. Just dropping the balloon is also fine—I just set $v_0 = 0$ and solve for the correct time to drop the balloon. The precision with which I must specify the velocity of the thrown balloon depends on the mathematical form of the equation, the specified values for the universal constant G, and the initial condition h_0. With the simple mathematical form of the equation and the actual gravity force constant G and height of the Leaning Tower of Pisa, hitting my friend should be relatively easy.

The three factors that are essential in predicting the motion of the water balloon illustrate the factors that are generally necessary to provide a purposeful outcome in engineering work: (1) the mathematical form that nature takes (as illustrated by the equation); (2) the values of the universal constants (G in the equation); and (3) the boundary conditions (which include the height h_0 of the tower from which I throw the balloon and the initial velocity v_0 with which I throw the balloon). The terms m and r may be thought of as additional boundary conditions that are specific to the location of the tower on the surface of the Earth (rather than some other location in the universe). The engineer has no control over the laws of nature and the mathematical form that they take. Neither does the engineer have any control over the universal constants such as the gravity force constant. The engineer can only set the boundary conditions, as when he draws up blueprints to specify exactly how a device will look when it has been fabricated.

Let us illustrate this design process with the requirements (or boundary conditions) that must be specified when an engineer designs an automobile. The engineer must very carefully prescribe the conditions under which the chemical energy in gasoline is released and converted to torque on the wheels of the car. Each dimension for each engine part is critical for the parts to work together harmoniously. The absolute size and shape of the parts (as distinct from the relative size to fit with each other) depend on the forces to be developed and transmitted, which in turn depend on the weight of the car and the speed it should achieve in service. The weight depends on the size, which in turn depends on the number of passengers plus luggage the car will carry. These factors then determine the size of the cylinders and pistons to be used in the engine and the rate of gasoline injected into these cylinders. The brake and suspension systems independently have to be scaled to fit the weight requirements, as do the specifications for the tires.

Notice how many of the specifications are related to each other and therefore cannot be independently specified or assigned. The greater this interdependence of specified boundary conditions, the more complex and demanding the design requirements. Small errors in the specification of any of these requirements will produce either a car with very inferior performance or, worse yet, a car that does not function at all.

Does the universe have these essential features that we associate with design? As it turns out, purposeful outcomes in the natural world depend on (1) the mathematical form that nature assumes; (2) the values of the universal constants; and (3) the initial or boundary conditions. While engineers can only fix the boundary conditions, the suitability of the universe as a habitat for life in general and for human beings in particular depends on all three. Thus, we will consider how each of these requirements appears to be essential in creating a suitable habitat for life.

A Remarkable Mathematical Form

Mathematics—in contrast to mere calculation—is an abstract intellectual activity that began in Greece in the sixth century B.C. Pythagoras was a key figure, as were his successors, Euclid and Archimedes. Their studies focused especially on geometric objects such as straight lines, circles, ellipses, and conic sections (i.e., the curves made by cutting a cone with a plane). In the third century B.C., Apollonias of Perga wrote eight monumental volumes devoted to these curves, describing their properties as "miraculous." Yet it never occurred to these mathematicians that such beautiful abstract forms from mathematics were descriptions of real phenomena. Imagine the delight of Johannes Kepler (1571–1630) some eighteen centuries later when he discovered that the orbits of planets around the sun conformed to these same beautiful but abstract mathematical forms. Kepler noted, "The chief aim of all investigations of the external world should be to discover the rational order and harmony which has been imposed on it by God and which He revealed to us in the language of mathematics."[2]

Galileo Galilei (1564–1642) observed that "the laws of nature are written by the hand of God in the language of mathematics." Morris Kline in his book *Mathematics: The Loss of Certainty*,[3] notes that the religious mathematicians of the sixteenth and seventeenth centuries—including Newton, Galileo, Kepler, and Copernicus—believed that the universe was orderly and thus described by mathematics because a rational God fashioned it that way. Kline says that these scientist-mathematicians believed that "God had designed the universe, and it was to be expected that all phenomena of nature would follow one master plan. One mind designing a universe would

160

almost surely have employed one set of basic principles to govern all related phenomena."[4]

The incredibly diverse phenomena we see in nature are characterized by a small number of physics laws, each of which assumes a simple mathematical form. Indeed, they can all be written on one side of one sheet of paper, as seen in Table 1.

Physicist Eugene Wigner, in a widely quoted paper titled "The Unreasonable Effectiveness of Mathematics in the Physical Sci-

Table 1

Fundamental Laws of Nature

Mechanics (Hamilton's Equations)

$$\dot{p} = -\frac{\partial H}{\partial q} \quad \dot{q} = \frac{\partial H}{\partial p}$$

Electrodynamics (Maxwell's Equations)

$$F^{\mu\nu} = \partial^\mu A^\nu - \partial^\nu A^\mu$$

$$\partial_\mu F^{\mu\nu} = j^\nu$$

Statistical Mechanics (Boltzmann's Equations)

$$S = -k\Sigma P_i \ln P_i$$

$$\frac{dS}{dt} \geq 0$$

Quantum Mechanics (Schrodinger's Equations)

$$i\hbar \,|\, \dot{\psi} \,\rangle = H \,|\, \psi \,\rangle$$

$$\Delta x \Delta p \geq \frac{\hbar}{2}$$

General Relativity (Einstein's Equation)

$$G\mu\nu = -8\pi \, GT_{\mu\nu}$$

ences,"[5] notes that scientists often take for granted the remarkable, even miraculous, effectiveness of mathematics in describing the real world. To quote Wigner, "The enormous usefulness of mathematics is something bordering on the mysterious....There is no rational explanation for it....The miracle of the appropriateness of the language of mathematics for the formulation of the laws of physics is a wonderful gift which we neither understand nor deserve. . . ."

Albert Einstein, in a letter to a friend, was struck by the mathematical comprehensibility of the world:

> You find it strange that I consider the comprehensibility of the world (to the extent that we are authorized to speak of such a comprehensibility) as a miracle or as an eternal mystery. Well, *a priori* one should expect a chaotic world which cannot be grasped by the mind in any way. . . . [T]he kind of order created by Newton's theory of gravitation, for example, is wholly different. Even if the axioms of the theory are proposed by man, the success of such a project presupposes a high degree of ordering of the objective world, and this could not be expected *a priori*. That is the "miracle" which is being constantly reinforced as our knowledge expands.[6]

Unlike Einstein and Wigner, but in keeping with Newton and his contemporaries, many modern physicists consider the remarkable mathematical form that nature assumes to be evidence for the existence of an intelligent designer. For example, the distinguished Russian physicist Alexander Polykov notes that, "We know that nature is described by the best of all possible mathematics because God created it." Australian astrophysicist Paul Davies says, "The equations of physics have in them incredible simplicity, elegance and beauty. That in itself is sufficient to prove to me that there must be a God who is responsible for these laws and responsible for the universe."

Nonetheless, mathematical form alone is insufficient to guarantee a universe that is a suitable habitat for life. The particular mathematical form is also critical. For example, it is essential that the mathematical form provide for stable systems at the atomic and cosmic level. The solutions to Hamilton's equations for non-relativistic Newtonian mechanics and for Einstein's theory of general relativity in Table 1 are unstable for a sun with planets unless the gravitational potential energy is proportional to r^{-1}, a requirement that is only met for a universe with three spatial dimensions.

For the solution to Schroedinger's equations (Table 1) for the hydrogen atom to give stable, bound energy levels, again, a universe with three (or fewer) spatial dimensions is required. Maxwell's equations (Table 1) are also only valid for a universe with three spatial dimensions. Furthermore, Richard Courant has found that high-fidelity transmission of electromagnetic or acoustic signals is optimized in our three-dimensional universe: "[O]ur actual physical world, in which acoustic or electromagnetic signals are the basis of communication, seems to be singled out among the mathematically conceivable models by intrinsic simplicity and harmony."[7]

In summary, it is clear that the specific mathematical character of our universe is essential for it to be a suitable habitat for life. Yet the reason that nature has this precise mathematical form is problematic from a naturalistic or materialistic worldview.

The Mystery of the Cosmological Constants

There are certain universal constants that are an essential part of our mathematical description of the universe. A partial list is found in Table 2 (p. 165) and includes Planck's constant h; the speed of light c; the gravity force constant G; the mass of the proton, electron, and neutron; the unit charge for the electron or proton; the weak force, strong nuclear force, and electromagnetic coupling constants; and Boltzmann's constant k.

When cosmological models were first developed in the midtwentieth century, it was naively assumed that the selection of a given set of constants was not critical to the formation of a suitable habitat for life. Subsequent parametric studies that systematically varied the constants have shown that changes in any of the constants produce a dramatically different universe that is unsuitable for life of any conceivable type.

Many books have appeared in the past ten years to summarize this surprising feature of our universe, to wit, that the universal constants have to be "just so" to have a universe suitable for life. A list of these books includes *The Anthropic Cosmological Principle*,[8] *Universes*,[9] *The Accidental Universe*,[10] *Superforce*,[11] *The Cosmic Blueprint*,[12] *Cosmic Coincidences*,[13] *The Anthropic Principle*,[14] *Universal Constants in Physics*,[15] *The Creation Hypothesis*,[16] and *Mere Creation*. I will illustrate this "just so" requirement for the various universal constants and properties of matter with several examples.

Physical Fine Structure Constants

The four forces in nature may each be expressed as dimensionless numbers to allow their relative strengths as they act in nature to be compared. These are summarized in Table 2 and vary by a factor of 10^{41} (10 with forty additional zeros after it), or equivalently by 41 orders of magnitude. Yet modest changes in any of these constants would produce dramatic changes in the universe and render it unsuitable for life. Several examples illustrate this fine-tuning of our universe.

Table 2 indicates that the electromagnetic force is 10^{38} times stronger than the gravity force. It is the force of gravity that draws protons together in stars, causing them to fuse together with a concurrent release of energy. The electromagnetic force causes them to repel. Because the gravity force is so weak compared to the electromagnetic force, the rate at which stars "burn" by fusion is very slow, allowing the stars to provide a stable source of energy over a very long period of time. If this ratio of strengths had been 10^{32} instead of 10^{38} (i.e., gravity were much stronger), stars would be a billion times less massive and would burn a million times faster.

The frequency distribution of electromagnetic radiation produced by the sun is also critical, as it needs to be tuned to the energies of chemical bonds on earth. If the photons of radiation are too energetic (too much ultraviolet radiation), then chemical bonds are destroyed and molecules are unstable; if the photons are too weak (too much infrared radiation), then chemical reactions will be too sluggish. The radiation produced is dependent on a careful balancing of the electromagnetic force (alpha-E) and the gravity force (alpha-G), with the mathematical relationship including (alpha-E)^{12}, making the specification for the electromagnetic force particularly critical. On the other hand, the chemical bonding energy comes from quantum mechanical calculations that include the electromagnetic force, the mass of the electron, and Planck's constant. Thus, all of these constants have to be sized relative to each other to give a universe in which radiation is tuned to the necessary chemical reactions that are essential for life.

Another fine-tuning coincidence is that the emission spectrum for the sun not only peaks at an energy level that is ideal to facilitate chemical reactions, but it also peaks in the optical window for water. Water is 10^7 times more opaque to ultraviolet and infrared radiation than it is to radiation in the visible spectrum (or what we call light). Since living tissue in general and eyes in particular are composed mainly of water, communication by sight would be impossible were it not for this unique window of light transmission by water being ideally matched to the radiation from the sun. Yet this matching requires carefully prescribing the values of the grav-

Table 2

Universal Constants

Boltzmann's constant	$k = 1.38 \times 10^{-23}$ j/°K
Planck's constant	$h = 6.63 \times 10^{-34}$ J/s
Speed of light	$c = 3.00 \times 10^8$ m/s
Gravitational constant	$G = 6.67 \times 10^{-11} \dfrac{N - m^2}{kg^2}$

Mass of Elementary Particles

Pion rest mass/energy	$m_\pi = 0.238 \times 10^{-24}$ kg/135 MeV
Neutron rest mass/energy	$m_n = 1.675 \times 10^{-27}$ kg/939.6 MeV
Electron rest mass	$m_e = 9.11 \times 10^{-31}$ kg/0.511 MeV
Proton rest mass	$m_p = 1.673 \times 10^{-27}$ kg/938.3 MeV
Unit charge	$e = 1.6 \times 10^{-19}$ coul
Mass-energy relation	$c^2 = \dfrac{E}{m}$ J/kg

Fine Structure Constants

Gravitation fine structure constant (alpha-G)

$$\alpha_g = [\frac{m_p^2}{hc} \cdot G] = 0.5 \times 10^{-40}$$

Fine structure constant of the weak interaction (alpha-W)

$$\alpha_w = [\frac{m_e^2 c}{\hbar^3} \cdot g_f] = 10^{-11}$$

Electromagnetic fine structure constant (alpha-E)

$$\alpha_e = [\frac{1}{\hbar c} e^2] = 1/137$$

Fine structure constant of the strong interaction (alpha-S)

$$\alpha_f = f = 3.9$$

ity and electromagnetic force constants, as well as Planck's constant and the mass of the electron.

Next consider the strength of the nuclear strong force. The most critical element in nature for the development of life is carbon. Yet, it has recently become apparent that the abundance of carbon in nature is the result of a very precise balancing of the strong force and the electromagnetic force, which determine the quantum energy levels for nuclei. Only certain energy levels are permitted for nuclei, and these may be thought of as steps on a ladder. If the mass–energy for two colliding particles results in a combined mass-energy that is equal to or *slightly less* than a permissible energy level on the quantum "energy ladder," then the two nuclei will readily stick together or fuse on collision, with the energy difference needed to reach the step being supplied by the kinetic energy of the colliding particles. If this mass-energy level for the combined particles is exactly right, or "just so," then the collisions are said to have resonance, which is to say that there is a high efficiency of collisions for fusing the colliding particles.

On the other hand, if the combined mass-energy results in a value that is *slightly higher* than one of the permissible energy levels on the energy ladder, then the particles will simply bounce off each other rather than stick together or fuse. In 1970 Fred Hoyle predicted the existence of the unknown resonance energy level for carbon, and he was subsequently proven right. The fusion of helium and beryllium give a mass-energy value that is 4 percent less than the resonance energy in carbon, which is easily made up by kinetic energy. Equally important was the discovery that the mass-energy for the fusion of carbon with helium was 1 percent greater than quantum energy level on the energy ladder for oxygen, making this reaction quite unfavorable. Thus, almost all beryllium is converted to carbon, but only a small fraction of the carbon is immediately converted to oxygen. These two results require the specification of the relative strength of the strong force and the electromagnetic force to within approximately 1 percent, which is truly remarkable given their large absolute values and difference of a factor of 100, as seen in Table 2.

More generally, a 2 percent increase in the strong force relative to the electromagnetic force leaves the universe with no hydrogen, no long-lived stars that burn hydrogen, and no water (which is a molecule composed of two hydrogen atoms and one oxygen atom), the ultimate solvent for life. A decrease of only 5 percent in the strong force relative to the electromagnetic force would prevent the formation of deuterons from combinations of protons and neutrons. This would in turn prevent the formation of all the heavier nuclei through fusion of deuterons to form helium, helium fusion with helium to form beryllium, and so forth. In 1980 Rozental estimated that the strong force had to be within 0.8 and 1.2 times its actual strength for there to be deuterons and all elements of atomic weight 4 or more.

If the weak force coupling constant (see Table 2) were slightly larger, neutrons would decay more rapidly, reducing the production of deuterons, and thus of helium and elements with heavier nuclei. On the other hand, if the weak force coupling constant were slightly weaker, the big bang would have burned almost all of the hydrogen into helium, with the ultimate outcome being a universe with little or no hydrogen and many heavier elements instead. This would leave no long-term stars and no hydrogen-containing compounds, especially water. In 1991 Breuer noted that the appropriate mix of hydrogen and helium to provide hydrogen-containing compounds, long-term stars, and heavier elements is approximately 75 percent hydrogen and 25 percent helium, which is just what we find in our universe.

This is only an illustrative and not an exhaustive list of cosmic coincidences. They clearly demonstrate how the four forces in nature have to be very carefully scaled to give a universe that provides long-term sources of energy and a variety of atomic building blocks necessary for life. Many other examples involving the fine-tuning of these forces are described in the books previously cited. Even so, the fine-tuning of the universe is not confined to these four forces. As it turns out, the elementary particles, as well as other universal constants like the speed of light and Planck's constant, also have to be very precisely specified.

Masses of Elementary Particles and Other Universal Constants

Scientists have been surprised to learn that the masses of the elementary particles must also be very carefully specified relative to each other and also to the forces in nature. For example, Stephen Hawking has noted that the difference in the mass of the neutron and the mass of the proton must be approximately equal to twice the mass of the electron. The mass-energy of the proton is 938.28 MeV, the mass-energy of the electron is 0.51 MeV, and the mass-energy of the neutron is 939.57 MeV. If the mass-energy of the proton plus the mass-energy of the electron were not slightly smaller than the mass-energy of the neutron, then electrons would combine with protons to form neutrons, with all atomic structure collapsing, leaving a world of neutrons only.

On the other hand, if this difference were larger, then neutrons would all decay into protons and electrons, leaving a world of hydrogen only, since neutrons are necessary for protons to combine to build heavier nuclei and the associated elements. As things are, the neutron is just heavy enough to ensure that the big bang would yield one neutron to every seven protons, allowing for an abundant supply of hydrogen for star fuel and enough neutrons to build up the heavier elements in the universe. Again, the precise relative values for the masses of these elementary particles are seen to be critical to provide a universe with long-term sources of energy and elemental diversity.

What are distinguished scientists saying about these cosmological coincidences? Freeman J. Dyson says, "As we look out into the universe and identify the many accidents of physics and astronomy that have worked to our benefit, it almost seems as if the universe must in some sense have known that we were coming."

Nobel laureate Arno Penzias makes this observation about the enigmatic character of the universe: "Astronomy leads us to a unique event, a universe which was created out of nothing and delicately balanced to provide exactly the conditions required to support life. In the absence of an absurdly-improbable accident, the observations of modern science seem to suggest an underlying, one might say, supernatural plan."

Sir Fred Hoyle, famous British astronomer who early on (1951) argued that the coincidences were just that, coincidences, by 1984 had changed his mind, as is evident from this quotation: "Such properties seem to run through the fabric of the natural world like a thread of happy coincidences. But there are so many odd coincidences essential to life that some explanation seems required to account for them."

The Remarkable Precision of Initial Conditions

The specific mathematical form that nature takes and the highly specific values of the various universal constants and masses of elementary particles by themselves cannot account for life. They are necessary but not sufficient conditions. All this elegant fine-tuning could have occurred as described above, and life still would not have occurred if the boundary conditions at certain critical points had not been properly set. We therefore turn to the initial conditions for the big bang.

A fundamental boundary condition of the big bang that is critical is its initial velocity. If this velocity is too fast, the matter in the universe expands too quickly and never coalesces into planets, stars, and galaxies. If the initial velocity is too slow, the universe expands only for a short time and then quickly collapses under the influence of gravity. Well-accepted cosmological models tell us that the initial velocity must be specified to a precision of $1/10^{55}$. This requirement seems to overwhelm chance and has been the impetus for creative alternatives, most recently the new inflationary model of the big bang.

Inflation itself, however, seems to require fine-tuning for it to occur at all and for it to yield irregularities neither too small nor too large for galaxies to form. Early on it was estimated that two components of an expansion-driving cosmological constant must cancel each other with an accuracy better than 1 part in 10^{50}. In the January 1999 issue of *Scientific American*,[17] the required accuracy was sharpened to 1 part in 10^{123}. Furthermore, the ratio of the gravitational energy to the kinetic energy must be equal to 1.00000 with a variation of 1 part in 100,000. Such estimates are being actively researched at the moment and their values may change over time. Nonetheless, it appears that very highly specified boundary conditions will be present in whatever model is finally confirmed for the big-bang origin of the universe.

No Cosmic Accident

My initial example of design was very simple. It involved one physical law, one universal constant, and two initial conditions. These could be prescribed so that my water balloon would arrive on the plaza of the Leaning Tower of Pisa just in time to hit my strolling friend. This is a relatively easy design problem.

However, for the universe to have stars that generate elemental diversity, provide long-term sources of energy of a suitable wavelength to facilitate chemical reactions, and satisfy many other requirements for a suitable habitat for life as well as for the origin of life, the mathematical form of the laws of nature, the nineteen universal constants (not all of which are listed in Table 2), and many initial conditions have to be *just so*.

Moreover, many of these requirements are interrelated. For instance, the initial velocity requirement is related to the strength of the gravity force. With so many distinct and interrelated requirements, it is difficult to imagine how all of these could have "accidentally" happened to be exactly what they needed to be. Given the many interdependent constraints, it appears unlikely that there is an alternative set of values for these constants that would work. Furthermore, the necessary values range over thirty orders of magnitude (10^{30}), making their accidentally correct "selection" all the more remarkable.

It is quite easy to understand why so many scientists have changed their minds in the past thirty years, agreeing that the universe cannot reasonably be explained as a cosmic accident. Evidence for an intelligent designer becomes more compelling the more we understand about our carefully crafted habitat.

SIGNS OF INTELLIGENCE

A Primer on the Discernment of Intelligent Design

WILLIAM A. DEMBSKI

William Dembski, Ph.D. (mathematics, University of Chicago, and phi-losophy, University of Illinois at Chicago), also holds an M.Div. from Princeton Theological Seminary. He is a fellow of the Discovery Insti-tute's Center for the Renewal of Science and Culture, the author of The Design Inference, *and* Intelligent Design: The Bridge Between Science and Theology,[1] *and editor of* Mere Creation.

*I*ntelligent design examines the distinction between three modes of explanation: necessity, chance, and design. In our worka-day lives we find it important to distinguish between these modes of explanation. Did she fall or was she pushed? And if she fell, was it simply bad luck or was her fall unavoidable? More generally, given an event, object, or structure, we want to know:

1. Did it have to happen?
2. Did it happen by accident?
3. Did an intelligent agent cause it to happen?

Given an event to be explained, the first thing to determine is whether it had to happen. If so, the event is necessary. By "necessary" I don't just mean logically necessary, as in true across all possible worlds, but I also include physical necessity, as in a law-like relation between antecedent circumstances and consequent events. Not all events are necessary.

Events that happen but do not have to happen are said to be contingent. In our everyday lives we distinguish two types of contingency: one blind, the other directed. A blind contingency lacks a superintending intelligence and is usually characterized by probabilities. Blind contingency is another name for chance. A directed contingency, on the other hand, is the result of a superintending intelligence. Directed contingency is another name for design.

An Ancient Question

This characterization of necessity, chance, and design is pretheoretical and therefore inadequate for building a precise scientific theory of design. We therefore need to inquire whether there is a principled way to distinguish these modes of explanation. Philosophers and scientists have disagreed not only about how to distinguish these modes of explanation, but also about their very legitimacy. The Epicureans, for instance, gave pride of place to chance. The Stoics, on the other hand, emphasized necessity and design, but rejected chance. In the Middle Ages, Moses Maimonides contended with the Islamic interpreters of Aristotle who viewed the heavens as, in Maimonides's words, "the necessary result of natural laws." Where the Islamic philosophers saw necessity, Maimonides saw design.

In arguing for design in his *Guide for the Perplexed*, Maimonides looked to the irregular distribution of stars in the heavens. For him that irregularity demonstrated contingency. But was that contingency the result of chance or design? Neither Maimonides nor the Islamic interpreters of Aristotle had any use for Epicurus and his views on chance. For them chance could never be fundamental but was at best a placeholder for ignorance. Thus for Maimonides and his Islamic colleagues, the question was whether a principled distinction could be drawn between necessity and design. Maimonides, arguing from observed contingency in nature, said yes. The Islamic philosophers, intent on keeping Aristotle pure of theology, said no.

A Modern Demise

Modern science has also struggled with how to distinguish between necessity, chance, and design. Newtonian mechanics, construed as a set of deterministic physical laws, seemed only to permit necessity. Nonetheless, in the General Scholium to his *Principia*, Newton claimed that the stability of the planetary system depended not only on the regular action of the universal law of gravitation, but also on the precise initial positioning of the planets and comets in relation to the sun. As he explained: "Though these bodies may, indeed, persevere in their orbits by the mere laws of gravity, yet they could by no means have at first derived the regular position of the orbits themselves from those laws.... [Thus] this most beautiful system of the sun, planets, and comets, could only proceed from the counsel and dominion of an intelligent and powerful being." Like Maimonides, Newton saw both necessity and design as legitimate explanations, but gave short shrift to chance.

Newton published his *Principia* in the seventeenth century. By the nineteenth century, necessity was still in, chance was still out, but design had lost much of its appeal. When asked by Napoleon where God fit into his equations of celestial mechanics, astronomer and mathematician Laplace famously replied, "Sire, I have no need of that hypothesis." In place of a designing intelligence that precisely positioned the heavenly bodies, Laplace proposed his nebular hypothesis, which accounted for the origin of the solar system strictly as the result of natural gravitational forces.

Since Laplace's day, science has largely dispensed with design. Certainly Darwin played a crucial role here by eliminating design from biology. Yet at the same time science was dispensing with design, it was also dispensing with Laplace's vision of a deterministic universe (recall Laplace's famous demon who could predict the future and retrodict the past with perfect precision provided that present positions and momenta of particles were fully known). With the rise of statistical mechanics and then quantum mechanics, the role of chance in physics came to be regarded as ineliminable. Consequently, a deterministic, necessitarian universe has given way to a stochastic universe in which chance and necessity are both regarded as fundamental modes of scientific explanation, neither being reducible to the other. To sum up, contemporary science allows a principled distinction between necessity and chance, but repudiates design.

Bacon and Aristotle

But was science right to repudiate design? My aim in *The Design Inference* is to rehabilitate design. I argue that design is a legitimate and fundamental mode of scientific explanation on a par with chance and necessity. Since my aim is to rehabilitate design, it will help to review why design was removed from science in the first place. Design, in the form of Aristotle's formal and final causes, after all, had once occupied a perfectly legitimate role within natural philosophy, or what we now call science. With the rise of modern science, however, these causes fell into disrepute.

We can see how this happened by considering Francis Bacon. Bacon, a contemporary of Galileo and Kepler, though himself not a scientist, was a terrific propagandist for science. Bacon was concerned about the proper conduct of science and provided detailed canons for experimental observation, the recording of data, and drawing inferences from data. What interests us here, however, is what he did with Aristotle's four causes. For Aristotle, to understand any phenomenon properly, one had to understand its four causes, namely its material, efficient, formal, and final cause.

Two points about Aristotle's causes are relevant to this discussion. First, Aristotle gave equal weight to all four causes and would have regarded any inquiry that omitted one of his causes as fundamentally deficient. Second, Bacon adamantly opposed the inclusion of formal and final causes within science (see his *Advancement of Learning*). For Bacon, formal and final causes belonged to metaphysics and not to science. Science, according to Bacon, needed to limit itself to material and efficient causes, thereby freeing science from the sterility that inevitably results when science and metaphysics are conflated. This was Bacon's line, and he argued it forcefully.

We see Bacon's line championed in our own day. For instance, in his book *Chance and Necessity*,[2] biologist and Nobel laureate Jacques Monod argued that chance and necessity alone suffice to account for every aspect of the universe. Now whatever else we might want to say about chance and necessity, they provide at best a reductive account of Aristotle's formal causes and leave no room for Aristotle's final causes. Indeed, Monod explicitly denies any place for purpose within science.

Now I don't want to give the impression that I'm advocating a return to Aristotle's theory of causation. There are problems with

Aristotle's theory, and it needed to be replaced. My concern, however, is with what replaced it. By limiting scientific inquiry to material and efficient causes, which are of course perfectly compatible with chance and necessity, Bacon championed a view of science that could only end up excluding design.

The Design Instinct

But suppose we lay aside *a priori* prohibitions against design. In that case, what is wrong with explaining something as designed by an intelligent agent? Certainly there are many everyday occurrences that we explain by appealing to design. Moreover, in our daily lives it is absolutely crucial to distinguish accident from design. We demand answers to such questions as: Did she fall or was she pushed? Did someone die accidentally or commit suicide? Was this song conceived independently or was it plagiarized? Did someone just get lucky on the stock market or was there insider trading?

Not only do we demand answers to such questions, but entire industries are also devoted to drawing the distinction between accident and design. Here we can include forensic science, intellectual property law, insurance claims investigation, cryptography, and random number generation—to name but a few. Science itself needs to draw this distinction to keep itself honest. As a January 1998 issue of *Science*[3] made clear, plagiarism and data falsification are far more common in science than we would like to admit. What keeps these abuses in check is our ability to detect them.

If design is so readily detectable outside of science, and if its detectability is one of the key factors keeping scientists honest, why should design be barred from the actual content of science? There's a worry here. The worry is that when we leave the constricted domain of human artifacts and enter the unbounded domain of scientific inquiry, the distinction between design and nondesign cannot be reliably drawn. Consider, for instance, the following remark by Darwin in the concluding chapter of his *Origin of Species*:

> Several eminent naturalists have of late published their belief that a multitude of reputed species in each genus are not real species; but that other species are real, that is, have been independently created. . . . Nevertheless they do not pretend that they can define, or even conjecture, which are the created forms of life, and which are those produced by secondary

laws. They admit variation as a *vera causa* in one case, they arbitrarily reject it in another, without assigning any distinction in the two cases.[4]

It's this worry of falsely attributing something to design (here construed as creation) only to have it overturned later, that has prevented design from entering science proper.

This worry, though perhaps understandable in the past, can no longer be justified. There does in fact exist a rigorous criterion for discriminating intelligently from unintelligently caused objects. Many special sciences already use this criterion, though in a pretheoretic form (e.g., forensic science, artificial intelligence, cryptography, archeology, and the search for extraterrestrial intelligence). In *The Design Inference* I identify and make precise this criterion. I call it the *complexity-specification criterion*. When intelligent agents act, they leave behind a characteristic trademark or signature—what I call *specified complexity*. The complexity-specification criterion detects design by identifying this trademark of designed objects.

The Complexity-Specification Criterion

A detailed explanation and justification of the complexity-specification criterion is technical and can be found in *The Design Inference*. Nevertheless, the basic idea is straightforward and easily illustrated. Consider how the radio astronomers in the movie *Contact* detected an extraterrestrial intelligence. This movie, based on a novel by Carl Sagan, was an enjoyable piece of propaganda for the SETI research program—the Search for Extraterrestrial Intelligence. To make the movie interesting, the SETI researchers in *Contact* actually did find an extraterrestrial intelligence (the nonfictional SETI program has yet to be so lucky).

How, then, did the SETI researchers in *Contact* convince themselves that they had found an extraterrestrial intelligence? To increase their chances of finding an extraterrestrial intelligence, SETI researchers monitor millions of radio signals from outer space. Many natural objects in space produce radio waves (e.g., pulsars). Looking for signs of design among all these naturally produced radio signals is like looking for a needle in a haystack. To sift through the haystack, SETI researchers run the signals they monitor through computers programmed with pattern-matchers. So long as a signal doesn't match

176

one of the preset patterns, it will pass through the pattern-matching sieve (and that even if it has an intelligent source). If, on the other hand, it does match one of these patterns, then, depending on the pattern matched, the SETI researchers may have cause for celebration.

The SETI researchers in *Contact* did find a signal worthy of celebration, namely the following:

```
110111011111011111110111111111101111111111111
111111111111111110111111111111111111011111111
111111111111111101111111111111111111111111111110
111111111111111111111111111110111111111111111
111111111111111111110111111111111111111111111
111111111111111111011111111111111111111111111
111111111111111101111111111111111111111111111
111111111111111111011111111111111111111111111
111111111111111111111111110111111111111111111
111111111111111111111111111111111111111110111
111111111111111111111111111111111111111111111
111111111111111111011111111111111111111111111
111111111111111111111111111111111111111111101
111111111111111111111111111111111111111111111
111111111111111111111111111011111111111111111
111111111111111111111111111111111111111111111
111111111111101111111111111111111111111111111
111111111111111111111111111111111111111111111
111111011111111111111111111111111111111111111
111111111111111111111111111111111111111111111
111101111111111111111111111111111111111111111
111111111111111111111111111111111111111111111
111111111011111111111111111111111111111111111
111111111111111111111111111111111111111111111
111111111111111111111
```

The SETI researchers in *Contact* received this signal as a sequence of 1,126 beats and pauses, where 1s correspond to beats and 0s to pauses. This sequence represents the prime numbers from 2 to 101, where a given prime number is represented by the corresponding number of beats (i.e., 1s), and the individual prime numbers are separated by pauses (i.e., 0s). The SETI researchers in *Contact* took this signal as decisive confirmation of an extraterrestrial intelligence.

177

What is it about this signal that implicates design? Whenever we infer design, we must establish three things: *contingency*, *complexity*, and *specification*. Contingency ensures that the object in question is not the result of an automatic and therefore unintelligent process that had no choice in its production. Complexity ensures that the object is not so simple that it can readily be explained by chance. Finally, specification ensures that the object exhibits the type of pattern characteristic of intelligence. Let us examine these three requirements more closely.

Contingency

In practice, to establish the contingency of an object, event, or structure, one must establish that it is compatible with the regularities involved in its production, but that these regularities also permit any number of alternatives to it. Typically these regularities are conceived as natural laws or algorithms. By being compatible with but not required by the regularities involved in its production, an object, event, or structure becomes irreducible to any underlying physical necessity. Michael Polanyi and Timothy Lenoir have both described this method of establishing contingency.

The method applies quite generally: the position of Scrabble pieces on a Scrabble board is irreducible to the natural laws governing the motion of Scrabble pieces; the configuration of ink on a sheet of paper is irreducible to the physics and chemistry of paper and ink; the sequencing of DNA bases is irreducible to the bonding affinities between the bases; and so on. In the case at hand, the sequence of 0s and 1s to form a sequence of prime numbers is irreducible to the laws of physics that govern the transmission of radio signals. We therefore regard the sequence as contingent.

Complexity

To see next why complexity is crucial for inferring design, consider the following sequence of bits:

110111011111

These are the first twelve bits in the previous sequence representing the prime numbers 2, 3, and 5 respectively. Now it is a sure

178

bet that no SETI researcher, if confronted with this twelve-bit sequence, is going to contact the science editor at the *New York Times*, hold a press conference, and announce that an extraterrestrial intelligence has been discovered. No headline is going to read, "Aliens Master First Three Prime Numbers!"

The problem is that this sequence is much too short (and thus too simple) to establish that an extraterrestrial intelligence with knowledge of prime numbers produced it. A randomly beating radio source might by chance just happen to produce this sequence. A sequence of 1,126 bits representing the prime numbers from 2 to 101, however, is a different story. Here the sequence is sufficiently long (and therefore sufficiently complex) to allow that an extraterrestrial intelligence could have produced it.

Complexity as I am describing it here is a form of probability. (Later in this essay I will require a more general conception of complexity to unpack the logic of design inferences. But for now complexity as a form of probability is all we need.) To see the connection between complexity and probability, consider a combination lock. The more possible combinations of the lock, the more complex the mechanism, and, correspondingly, the more improbable that the mechanism can be opened by chance. Complexity and probability therefore vary inversely: the greater the complexity, the smaller the probability. Thus to determine whether something is sufficiently complex to warrant a design inference is to determine whether it has sufficiently small probability.

Even so, complexity (or improbability) isn't enough to eliminate chance and establish design. If I flip a coin one thousand times, I'll participate in a highly complex (i.e., highly improbable) event. Indeed, the sequence I end up flipping will be one in a trillion trillion trillion. . ., where the ellipsis indicates twenty-two more "trillions." This sequence of coin tosses won't, however, trigger a design inference. Though complex, this sequence won't exhibit a suitable pattern. Contrast this with the previous sequence representing the prime numbers from 2 to 101. Not only is this sequence complex, but it also embodies a suitable pattern. The SETI researcher who in the movie *Contact* discovered this sequence put it this way: "This isn't noise; this has structure."

Specification

What is a *suitable* pattern for inferring design? Not just any pattern will do. Some patterns can legitimately be employed to infer design whereas others cannot. The intuition underlying the distinction between patterns that alternately succeed or fail to implicate design is, however, easily motivated. Consider the case of an archer. Suppose an archer stands fifty meters from a large wall with bow and arrow in hand. The wall is sufficiently large that the archer cannot help but hit it. Now suppose each time the archer shoots an arrow at the wall, the archer paints a target around the arrow so that the arrow sits squarely in the bull's-eye. What can be concluded from this scenario? Absolutely nothing about the archer's ability as an archer. Yes, a pattern is being matched, but it is a pattern fixed only after the arrow has been shot. The pattern is thus purely *ad hoc*.

But suppose instead the archer paints a fixed target on the wall and then shoots at it. Suppose the archer shoots a hundred arrows, and each time hits a perfect bull's-eye. What can be concluded from this second scenario? Confronted with this second scenario we are obligated to infer that here is a world-class archer, one whose shots cannot legitimately be referred to luck, but rather must be referred to the archer's skill and mastery. Skill and mastery are of course instances of design.

The archer example introduces three elements that are essential for inferring design:

1. A reference class of possible events (here the arrow hitting the wall at some unspecified place);
2. A pattern that restricts the reference class of possible events (here a target on the wall); and
3. The precise event that has occurred (here the arrow hitting the wall at some precise location).

In a design inference, the reference class, the pattern, and the event are linked, with the pattern mediating between event and reference class, and helping to decide whether the event is due to chance or design. Note that in determining whether an event is sufficiently improbable or complex to implicate design, the relevant improbability is not that of the precise event that occurred, but that of the target/pattern. Indeed, the bigger the target, the easier it is to hit it by chance and thus apart from design.

180

The type of pattern in which an archer fixes a target first and then shoots at it is common to statistics, where it is known as setting a *rejection region* prior to an experiment. In statistics, if the outcome of an experiment falls within a rejection region, the chance hypothesis supposedly responsible for the outcome is rejected. The reason for setting a rejection region prior to an experiment is to forestall what statisticians call "data snooping" or "cherry picking." Just about any data set will contain strange and improbable patterns if we look hard enough. By forcing experimenters to set their rejection regions prior to an experiment, the statistician protects the experiment from spurious patterns that could just as well result from chance.

Now a little reflection makes clear that a pattern need not be given prior to an event to eliminate chance and implicate design. Consider the following cipher text:

nfuijolt ju jt mjlf b xfbtfm

Initially this looks like a random sequence of letters and spaces— you lack any pattern for rejecting chance and inferring design.

But suppose that someone comes along and tells you to treat this sequence as a Caesar cipher, moving each letter one notch down the alphabet. Now the sequence reads,

methinks it is like a weasel

Even though the pattern (in this case, the decrypted text) is given after the fact, it still is the right sort of pattern for eliminating chance and inferring design. In contrast to statistics, which always identifies its patterns before an experiment is performed, cryptanalysis must discover its patterns after the fact. In both instances, however, the patterns are suitable for inferring design.

Patterns thus divide into two types: those that in the presence of complexity warrant a design inference and those that, despite the presence of complexity, do not warrant a design inference. The first type of pattern I call a *specification*, the second a *fabrication*. Specifications are the non-*ad hoc* patterns that can legitimately be used to eliminate chance and warrant a design inference. In contrast, fabrications are the *ad hoc* patterns that cannot legitimately be used to warrant a design inference.

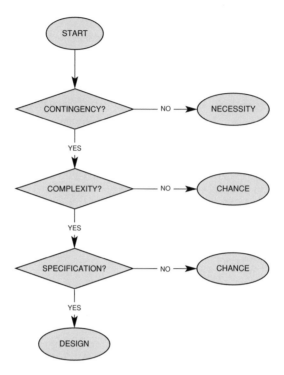

To sum up, the complexity-specification criterion detects design by establishing three things: contingency, complexity, and specification. When called to explain an event, object, or structure, we have to decide: Are we going to attribute it to *necessity*, *chance*, or *design*? According to the complexity-specification criterion, to answer this question is to answer three simpler questions: Is it contingent? Is it complex? Is it specified? Consequently, the complexity-specification criterion can be represented as a flowchart with three decision nodes. I call this flowchart the Explanatory Filter.

Independent Patterns Are Detachable

For a pattern to count as a specification, the important thing is not when it was identified, but whether in a certain well-defined sense it is *independent* of the event it describes. Drawing a target around an arrow already embedded in a wall is not independent of the arrow's trajectory. Consequently, such a target/pattern cannot be used to attribute the arrow's trajectory to design. Patterns that are specifications cannot simply be read off the events whose design is in question. Rather, to count as specifications, patterns must be suitably independent of events. I refer to this relation of independence as *detachability*, and say that a pattern is detachable only if it satisfies that relation.

Detachability can be understood as asking this question: Given an event (whose design is in question) and a pattern describing it, would we be able to construct that pattern if we had no knowledge of which event occurred? Assume an event has occurred. A pattern describing the event is given. The event is one from a range of possible events. If all we knew was the range of possible events without any specifics about which event actually occurred, could we still construct the pattern describing the event? If so, the pattern is detachable from the event.

A Trick with Coins

To see what's at stake, consider the following example. (It was this example that finally clarified for me what transforms a pattern *simpliciter* into a pattern *qua* specification.) The following event E to all appearances was obtained by flipping a fair coin one hundred times:

THTTTHHTHHTTTTTHTHTTHHHTTHTHHHTH
HTTTTTTTHTTHTTTHHTHTTTHTHTHHTTHH
HTTTHTTHHTHTHTHHHHTTHHTHHHHTHHH
HTT   E

Is E the product of chance or not? A standard trick of statistics professors with an introductory statistics class is to divide the class in two and have students in one half of the class each flip a coin one hundred times and write down the sequence of heads and tails on a slip of paper; students in the other half each generate with their minds a "random-looking" string that mimics the tossing of a coin one hundred times and also write down the sequence of heads and tails on a slip of paper. When the students then hand in their slips of paper, it is the professor's job to sort the papers into two piles, those generated by flipping a fair coin, and those concocted in the students' heads. To the amazement of the students, the statistics professor is typically able to sort the papers with 100 percent accuracy.

183

There's no mystery here. The statistics professor simply looks for a repetition of six or seven heads or tails in a row to distinguish the truly random from the pseudo-random sequences. In a hundred coin flips, one is quite likely to see such a repetition. On the other hand, people concocting pseudo-random sequences with their minds tend to alternate between heads and tails too frequently. Whereas with a truly random sequence of coin tosses there is a 50 percent chance that one toss will differ from the next, as a matter of human psychology people expect that one toss will differ from the next around 70 percent of the time.

How, then, will our statistics professor fare when confronted with E? Will she attribute E to chance or to the musings of someone trying to mimic chance? According to the professor's crude randomness checker, E would be assigned to the pile of sequences presumed to be truly random, for E contains a repetition of seven tails in a row. Everything that at first blush would lead us to regard E as truly random checks out. There are exactly fifty alternations between heads and tails (as opposed to the seventy that would be expected from human beings trying to mimic chance). What's more, the relative frequencies of heads and tails check out: there were forty-nine heads and fifty-one tails. Thus it's not as though the coin supposedly responsible for generating E was heavily biased in favor of one side versus the other.

But Is It Really Chance?

Suppose, however, that our statistics professor suspects she is not up against a neophyte statistics student, but instead a fellow statistician who is trying to put one over on her. To help organize her problem, study it more carefully, and enter it into a computer, she will find it convenient to let strings of 0s and 1s represent the outcomes of coin flips, with 1 corresponding to heads and 0 to tails. In that case the following pattern D will correspond to the event E:

0100011011000001010011100101110111000000010010
0011010001010110011110001001101010111100110111
10111100   D

Now, the mere fact that the event E conforms to the pattern D is no reason to think that E did not occur by chance. As things stand, the pattern D has simply been read off the event E.

But D need not have been read off of E. Indeed, D could have been constructed without recourse to E. To see this, let us rewrite D as follows:

<pre>
         0
         1
         00
         01
         10
         11
         000
         001
         010
         011
         100
         101
         110
         111
         0000
         0001
         0010
         0011
         0100
         0101
         0110
         0111
         1000
         1001
         1010
         1011
         1100
         1101
         1110
         1111
         00   D
</pre>

By viewing D this way, anyone with the least exposure to binary arithmetic immediately recognizes that D was constructed simply by writing binary numbers in ascending order, starting with the one-digit binary numbers (i.e., 0 and 1), proceeding then to the two-digit binary numbers (i.e., 00, 01, 10, and 11), and continuing on until one hundred digits were recorded. It's therefore intuitively clear that D does not describe a truly random event (i.e., an event gotten by tossing a fair coin), but rather a pseudo-random event, concocted by doing a little binary arithmetic.

Side Information Does the Trick

Although it's now intuitively clear why chance cannot properly explain E, we need to consider more closely why this mode of explanation fails here. We started with a putative chance event E, supposedly the result of flipping a fair coin one hundred times. Since heads and tails each have probability $\frac{1}{2}$, and since this probability gets multiplied for each flip of the coin, it follows that the probability of E is 2^{-100}, or approximately 10^{-30}.

In addition, we constructed a pattern D to which E conforms. Initially D proved insufficient to eliminate chance as the explanation of E since in its construction D was simply read off E. Rather, to eliminate chance we also had to recognize that D could have been constructed quite easily by performing some simple arithmetic operations with binary numbers. Thus to eliminate chance we needed to employ additional *side information*, which in this case consisted of our knowledge of binary arithmetic. This side information detached the pattern D from the event E and thereby rendered D a specification.

For side information to detach a pattern from an event, it must satisfy two conditions, *conditional independence* and *tractability*. First, the side information must be conditionally independent of the event E. Conditional independence, a well-defined notion from probability theory, means that the probability of E doesn't change once the side information is taken into account. Conditional independence is the standard probabilistic way of unpacking epistemic independence. Two things are epistemically independent if knowledge about one thing (in this case the side information) does not affect knowledge about the other (in this case the occurrence of E). This is certainly the case here since our knowledge of binary arithmetic does not affect the probabilities we assign to coin tosses.

The second condition, the tractability condition, requires that the side information enable us to construct the pattern D to which E conforms. This is evidently the case here as well, since our knowledge of binary arithmetic enables us to arrange binary numbers in ascending order, and thereby construct the pattern D.

But what exactly is this ability to construct a pattern on the basis of side information? Perhaps the most slippery words in philosophy are "can," "able," and "enable." Fortunately, just as there is a precise theory for characterizing the epistemic independence between an event and side information—namely, probability theory—so too there is a precise theory for characterizing the ability to construct a pattern on the basis of side information—namely, complexity theory.

Complexity Theory

Complexity theory, conceived now quite generally and not merely as a form of probability, assesses the difficulty of tasks given the resources available for accomplishing those tasks. If I may generalize computational complexity theory, it ranks tasks according to difficulty and then determines which tasks are sufficiently manageable to be doable or tractable. For instance, given current technology we find sending a person to the moon tractable, but sending a person to the nearest galaxy intractable.

In the tractability condition, the task to be accomplished is the construction of a pattern, and the resources for accomplishing that task are side information. Thus, for the tractability condition to be satisfied, side information must provide the resources necessary for constructing the pattern in question. All of this admits a precise complexity-theoretic formulation and makes definite what I called "the ability to construct a pattern on the basis of side information."

Taken jointly, the tractability and conditional independence conditions mean that side information enables us to construct the pattern to which an event conforms, yet without recourse to the actual event. This is the crucial insight. Because the side information is conditionally and therefore epistemically independent of the event, any pattern constructed from this side information is obtained without recourse to the event. In this way any pattern that is constructed

187

from such side information avoids the charge of being *ad hoc*. These, then, are the detachable patterns. These are the specifications.

A Matter of Choice

The complexity-specification criterion is exactly the right instrument for detecting design. To see why, we need to understand what makes intelligent agents detectable in the first place. The principal characteristic of intelligent agency is choice. Even the etymology of the word "intelligent" makes this clear. "Intelligent" derives from two Latin words, the preposition *inter*, meaning between, and the verb *lego*, meaning to choose or select. Thus, according to its etymology, intelligence consists in *choosing between*. For an intelligent agent to act is therefore to choose from a range of competing possibilities.

This is true not just of humans, but of animals as well as of extraterrestrial intelligences. A rat navigating a maze must choose whether to go right or left at various points in the maze. When SETI researchers attempt to discover intelligence in the extraterrestrial radio transmissions they are monitoring, they assume an extraterrestrial intelligence could have chosen any number of possible radio transmissions, and then attempt to match the transmissions they observe with certain patterns as opposed to others. Whenever a human being utters meaningful speech, a choice is made from a range of possible sound combinations that might have been uttered. Intelligent agency always entails discrimination, choosing certain things, ruling out others.

Recognizing Intelligence

Given this characterization of intelligent agency, the crucial question is how to recognize it. Intelligent agents act by making a choice. How, then, do we recognize that an intelligent agent has made a choice? A bottle of ink spills accidentally onto a sheet of paper; someone takes a fountain pen and writes a message on a sheet of paper. In both instances ink is applied to paper. In both instances one among an almost infinite set of possibilities is realized. In both instances a contingency is actualized and others are ruled out. Yet in one instance we ascribe agency, in the other chance.

188

What is the relevant difference? Not only do we need to observe that a contingency was actualized, but we need also to be able to specify that contingency. The contingency must conform to an independently given pattern, and we must be able independently to construct that pattern. A random inkblot is unspecified; a message written with ink on paper is specified. The exact message recorded may not be specified, but orthographic, syntactic, and semantic constraints will nonetheless specify it.

Actualizing one among several competing possibilities, ruling out the rest, and specifying the one that was actualized encapsulates how we recognize intelligent agency, or equivalently, how we detect design. Experimental psychologists who study animal learning and behavior have known this all along. To learn a task an animal must acquire the ability to actualize behaviors suitable for the task as well as the ability to rule out behaviors unsuitable for the task. Moreover, for a psychologist to recognize that an animal has learned a task, it is necessary not only to observe the animal making the appropriate discrimination, but also to specify the discrimination.

Rats and Mazes

Thus, to recognize whether a rat has successfully learned how to traverse a maze, a psychologist must first specify which sequence of right and left turns conducts the rat out of the maze. No doubt, a rat randomly wandering a maze also discriminates a sequence of right and left turns. But by randomly wandering the maze, the rat gives no indication that it can discriminate the appropriate sequence of right and left turns for exiting the maze. Consequently, the psychologist studying the rat will have no reason to think the rat has learned how to traverse the maze.

Only if the rat executes the sequence of right and left turns specified by the psychologist will the psychologist recognize that the rat has learned how to traverse the maze. Now it is precisely the learned behaviors we regard as intelligent in animals. Hence it is no surprise that the same scheme for recognizing animal learning recurs for recognizing intelligent agency generally, to wit: actualizing one among several competing possibilities, ruling out the others, and specifying the one actualized.

Note that complexity is implicit here as well. To see this, consider again a rat traversing a maze, but now take a very simple maze in which two right turns conduct the rat out of the maze. How will a psychologist studying the rat determine whether it has learned to exit the maze? Just putting the rat in the maze will not be enough. Because the maze is so simple, the rat by chance could just happen to take two right turns and thereby exit the maze. The psychologist will therefore be uncertain whether the rat actually learned to exit this maze or just got lucky.

But contrast this with a complicated maze in which a rat must take just the right sequence of left and right turns to exit the maze. Suppose the rat must take one hundred appropriate right and left turns, and that any mistake will prevent the rat from exiting the maze. A psychologist who sees the rat take no erroneous turns and quickly exit the maze will be convinced that the rat has indeed learned how to exit the maze, and that it was not dumb luck.

This general scheme for recognizing intelligent agency is but a thinly disguised form of the complexity-specification criterion. In general, to recognize intelligent agency we must observe an actualization of one among several competing possibilities, note which possibilities were ruled out, and then be able to specify the possibility that was actualized. What's more, the competing possibilities that were ruled out must be live possibilities, and sufficiently numerous so that specifying the possibility that was actualized cannot be attributed to chance. In terms of complexity, this is just another way of saying that the range of possibilities is complex. In terms of probability, this is just another way of saying that the possibility that was actualized has small probability.

All the elements in this general scheme for recognizing intelligent agency (i.e., actualizing, ruling out, and specifying) find their counterpart in the complexity-specification criterion. It follows that this criterion formalizes what we have been doing right along when we recognize intelligent agency. The complexity-specification criterion pinpoints how we detect design.

Design, Metaphysics, and Beyond

Where is this work on design heading? Specified complexity, that key trademark of design, is, as it turns out, a form of information

190

(though one considerably richer than Claude Shannon's purely statistical form of it). Although called by different names and developed with different degrees of rigor, specified complexity is starting to have an effect on the special sciences.

For instance, specified complexity is what Michael Behe has uncovered with his irreducibly complex biochemical machines, what Manfred Eigen regards as the great mystery of life's origin, what for cosmologists underlies the fine-tuning of the universe, what David Chalmers hopes will ground a comprehensive theory of human consciousness, what enables Maxwell's demon to outsmart a thermodynamic system tending toward thermal equilibrium, and what within the Kolmogorov-Chaitin theory of algorithmic information identifies the highly compressible, nonrandom strings of digits. How complex specified information gets from an organism's environment into an organism's genome was one of the key questions at the October 1999 Santa Fe Institute symposium, "Complexity, Information & Design: A Critical Appraisal."

Shannon's purely statistical theory of information is giving way to a richer theory of complex specified information whose possibilities are only now coming to light. A natural sequel to *The Design Inference* is therefore to develop a general theory of complex specified information.

Yet despite its far-reaching implications for science, I regard the ultimate significance of this work on design to lie in metaphysics. In my view, design died not at the hands of nineteenth-century evolutionary biology, but at the hands of the mechanical philosophy two centuries earlier—and that despite the popularity of British natural theology at the time. Though the originators of the mechanical philosophy were typically theists, the design they retained was at best an uneasy rider on top of a mechanistic view of nature. Design is neither use nor ornament within a strictly mechanistic world of particles or other mindless entities organized by equally mindless principles of association, even if these be natural laws ordained by God.

The primary challenge, once the broader implications of design for science have been worked out, is therefore to develop a relational ontology in which the problem of being resolves thus: to be is to be in communion, and to be in communion is to transmit and receive information. Such an ontology will not only safeguard science and leave adequate breathing space for design, but will also make sense of the world as sacrament.

The world is a mirror representing the divine life. The mechanical philosophy was ever blind to this fact. Intelligent design, on the other hand, readily embraces the sacramental nature of physical reality. Indeed, intelligent design is just the Logos theology of John's Gospel restated in the idiom of information theory.

14

IS INTELLIGENT DESIGN SCIENCE?

The Scientific Status and Future
of Design-Theoretic Explanations

BRUCE L. GORDON

Bruce L. Gordon, Ph.D. (history and philosophy of physics, Northwestern University) is the interim director of the Baylor University's Science and Religion Project and an assistant research professor in Baylor's Institute of Faith and Learning. He was recently a postdoctoral fellow at the Center for Philosophy of Religion at the University of Notre Dame, and is presently at work on a series of articles leading to a book on the metaphysical import of quantum statistics.

The subject of this essay falls largely within the domain of the philosophy of science, so some general remarks about the philosophy of science and its importance are in order. But first a caveat: a lot of territory is going to be covered in a few pages here, and in the process, many of the subtleties of various issues will be lost. Nevertheless, it is still possible to convey the core of the necessary arguments, along with a few helpful illustrations, without distortion.

What Is the Philosophy of Science, and Why Is It Important?

Science plays an important role in contemporary culture. It is often held up as the pinnacle of human achievement, the standard against which all claims to knowledge must be compared, and the rule to which they must yield if found in conflict. Since it so permeates our lives, it is important for us to reflect upon the significance of science both as an activity and as a mode of knowledge acquisition. This is a central task of the *philosophy of science,* the branch of study that critically reflects on the nature of science and on how science functions in terms of its principles and practices.

Important questions in the philosophy of science include: What is the nature of science? Does science even have a nature? Is it possible to distinguish between genuine science and pseudo-science, and what would be the criteria for such a distinction? What are the goals and methodologies of science, and are they general or contextual? What are the presuppositions and limitations of scientific inquiry? Must the methodology of science be thoroughly naturalistic? What are "laws of nature"? What constitutes a scientific explanation of a phenomenon? How are scientific explanations tested? How do scientists go about justifying their claims? How does a theory gain acceptance in the scientific community? How do the historical and cultural settings in which scientific work is done affect its results? What is scientific progress? How does scientific knowledge advance and grow? Beyond these universal concerns, there are also questions that can be asked regarding the conceptual foundations of individual theories in physics and biology. Quantum mechanics, relativity, and the different branches of evolutionary biology all generate issues of interpretation and significance that are specific to their contexts, and serve as the basis for further specialization within the philosophy of science itself.

But such questions usually are not raised in the course of scientific research. For example, a particle physicist might concern himself with the explanation of a peculiar scattering cross-section from an accelerator experiment, but never inquire whether there are general features a scientific explanation must have if it is to be acceptable. A biologist might perform experiments with fruit flies to determine how heredity works, but not reflect at length on the relationship between observations and theories. As a consequence, many practicing scientists have an idealized conception of what science is and how it functions that does not do justice to actual scientific practice. While modern science embodies an impressive collection of intellectual achievements, and while the importance of science in our culture cannot be over-emphasized, the status of its conclusions and the objectivity of its methodology are not as clear-cut as many believe. For this reason, it is important to reflect on the nature of scientific explanations.

One metaphysical assumption undergirding much contemporary science is that the universe in its origin and function is a closed system of undirected physical processes. While many scientists reject this metaphysical picture, they still think that it is essential for science to function as if it were true. This means that they have accepted methodological naturalism as a necessary constraint on their practice as scientists. *Methodological naturalism* is the doctrine that says that for an explanation to be scientific it must be naturalistic, that is, it must only appeal to entities, causes, events, and processes contained within the material universe.

But is this restriction on scientific explanations necessary? And what might the practice of science look like if it were sometimes permissible to suspend this methodological assumption?

Scientific Theories and Explanations

The questions that philosophers of science think about are complex and difficult, so it is not surprising that they frequently disagree on the answers. The attempt to characterize scientific theories and scientific explanations is no exception.

Judge Overton's Criteria

A recent court case takes us to the heart of the issue. In the early 1980s there was a controversy in Arkansas over the teaching of "creation science." The state had mandated that creation science be given equal time in the classroom with the theory of evolution. Creation science proponents argue that the earth is no more than ten thousand years old, that all fossilization has been the product of a catastrophic worldwide flood (at the time of Noah), and that species have not evolved, but are each the direct product of God's creative acts. The constitutionality of the Arkansas statute was immediately challenged in the courts by a variety of groups.

One of the issues central to the discussion was whether creation science was properly scientific. To deal with this issue, the court had to answer at least three questions:

(a) What is science?
(b) When is a claim scientific?
(c) How do we distinguish science from nonscience or pseudo-science?

In his decision overturning the statute, Judge William R. Overton concluded that creation science was not a scientific theory at all, and should not be given equal time in the schools. On the basis of the testimony of several philosophers of science called in as expert witnesses, Overton defined a scientific theory as having the following features:

1. It is guided by natural law;
2. It is explanatory by reference to natural law;
3. It is testable against the empirical world;
4. Its conclusions are tentative, i.e., they are not necessarily the final word; and
5. It is falsifiable.

Overton's statement gives a good sense of what many scientists regard as the primary characteristics of the sort of explanations for which they search. But most philosophers of science tended to think that this distilled "essence" of science was a put-up job, pushed through the courts as part of a political agenda, and was not to be taken seriously as a characterization of scientific practice. It will help to understand why they thought this.

A number of philosophers of science in this century have attempted to give an account of what it means to offer a scientific explanation for a phenomenon. We will briefly consider three such accounts: the *deductive-nomological model,* the *causal-statistical* (statistical-relevance) *model,* and the *pragmatic model.* By considering these three, we will see where the problems lie and why Overton's characterization is inadequate.

196

The D-N Model

The *deductive-nomological* (D-N) *model* was the earliest formal model of scientific explanation. It has been very influential, and has much in common with Overton's account. It postulates four criteria for scientific explanations:

1. The thing to be explained (the *explanandum*) must be a logical consequence of the explanation offered (the *explanans*). In other words, the explanation must be able to be put in the form of a valid deductive argument, with the *explanandum* as its conclusion.
2. The *explanans* must contain at least one general law, and it must actually be required for the derivation of the *explanandum*, that is, if the law(s) were deleted from the explanation without adding any new premises, the explanation would no longer be valid.
3. The *explanans* must have empirical content, that is, it must be capable, at least in principle, of being tested.
4. The sentences constituting the *explanans* must be true.

Subsequently, it became clear that the D-N model had irremediable shortcomings falling into two categories: (a) there are arguments meeting the criteria of the D-N model that fail to be genuine scientific explanations; and (b) there are genuine scientific explanations that fail to meet the criteria of the D-N model. In short, these four criteria are neither sufficient nor necessary to guarantee that an explanation is scientific. To see this, consider three standard counterexamples to the model: the flagpole and its shadow, the man and the Pill, and syphilis and paresis (see Brittan and Lambert,[1] and Salmon *et al*[2] for a more extensive discussion).

That the D-N model is insufficient as an account of scientific explanation can be illustrated by considering the shadow cast by a flagpole. The length of the shadow is explained by the height of the pole and the angle of inclination of the sun. This explanation can be put into D-N form using the law that light takes the path of shortest distance through space (what we would judge to be a straight line). But note also that we can deduce the height of the flagpole from the length of its shadow and the elevation of the sun, and the elevation of the sun from the height of the flagpole and the length of its shadow. But this is problematic because the proper explanation for the height of the flagpole lies in its construction, and the proper explanation for the elevation of the sun lies in the geographical location of the observer, the season of the year, and the time of day. In order for the D-N model to have legitimate application, the question of particular causation needs to be taken into account. Causes are not properly explained in terms of their effects, e.g., the length of the shadow does not cause the sun to be at a certain elevation.

Consider this humorous counterexample. A man explains his failure to become pregnant over the last year, despite an amorous relationship with his wife, on the grounds that he has regularly consumed her birth control pills. He appeals to the law-like generalization that every man who regularly takes oral contraceptives will not get pregnant. This example conforms to the D-N pattern of explanation. The problem is that the birth control pills are irrelevant because men do not get pregnant. So it is possible to construct valid arguments with true premises in which some fact asserted by the premises is irrelevant to the real explanation of the phenomenon in question. Both the flagpole and the pregnancy examples speak to the insufficiency of the D-N model as a guarantee of proper scientific explanation.

To see that the model does not provide conditions that are necessary for a proper scientific explanation, consider the explanation for the development of paresis (a form of tertiary syphilis characterized by progressive physical paralysis and loss of mental function). In order to develop paresis, it is necessary to have untreated latent syphilis, but only about 25 percent of the people in this situation ever develop it. So we have a necessary condition for the development of the disease, but we cannot use this to derive the conclusion that paresis will develop in an individual case, or even to predict that it will. In fact, we're better off predicting that it will not develop, since it doesn't in 75 percent of the cases. Even so, the proper scientific explanation for paresis is that it results from untreated latent syphilis. This is just one example of a scientific explanation which does not conform to the D-N model.

The Causal-Statistical Model

To remedy the defects of the D-N model of scientific explanation, the *causal-statistical* or *statistical-relevance model* was proposed. Advocates of this model stress the role of *causal* components in scientific explanations and generally deny that explaining something scientifically *must* involve rigorous deductive or inductive arguments. Because they recognize that there are rational explanations for *unexpected* events (like the onset of paresis after untreated latent syphilis), they reject the idea that universal or statistical laws and empirical facts must provide conditions of adequacy for a scientific explanation for the occurrence of events. Consequently, they downplay the idea that scientific explanations must be structured as an argument.

The positive idea behind the causal-statistical model is that a scientific explanation presents two things: (1) the set of factors statistically relevant to the occurrence of that event; and (2) the causal framework or link connecting those factors with the event to be explained. *Statistical relevance* may be defined as follows: factor B is *statistically relevant* to factor A if and only if the probability of A, given that B has already occurred, is different from the probability of A occurring on its own, that is, $P(A \mid B) \neq P(A)$. The causal network or link connecting the factors with an event is simply an account of the underlying causal processes and interactions that bring it about. A *causal process* is a continuous spatial and temporal process; a *causal interaction* is a relatively brief event in which two or more causal processes intersect. The causal-statistical theory arose from the conviction that legitimate scientific explanations have to explain events in terms of the things that actually caused them to happen.

While the causal-statistical model seems fairly solid, it nonetheless finds a counter-example in *quantum mechanics*, the theory describing the behavior of atomic and subatomic particles. The details of why the model fails for quantum mechanics are complicated. Roughly, the causal-statistical account appeals to processes that are deterministic and continuous in space and time, while it is generally accepted that quantum mechanics is not consistent with this view of the world. Since quantum mechanics is regarded as one of the triumphs of twentieth-century science and it is inconsistent with the causal-statistical account, we have a compelling reason for thinking that this model of explanation is too narrow. So an explanation does not have to conform to the causal-statistical criteria in order to be scientific.

The Pragmatic Model

The shortcomings of the D-N and causal-statistical models led to a third proposal for scientific explanations, the *pragmatic model*. This model has been defended most recently (and ably) by Bas van

Fraassen, a philosopher of science at Princeton University. Van Fraassen not only denies that scientific explanations have a characteristic form (as in the D-N model), but he also denies that they supply distinctive information (as in the causal–statistical model) outside of that provided by the theories, facts, and procedures of science itself. Calling an explanation "scientific" means nothing more than saying it draws on what we call science to provide an explanation, and whether this criterion is satisfied is something determined by the community of scientists themselves.

Beyond this, scientific explanations do not have any *essential* characteristics. For this reason, van Fraassen is skeptical of any essential distinction between science and nonscience, or science and pseudoscience. Finally, he rejects the idea that laws, by way of deduction, induction, or causal networks, are a necessary component of every scientific explanation. This does not mean that he rejects nomological explanations when they are appropriate. It just means that he rejects the idea that all genuine scientific explanations must be forced into a particular mold.

On the positive side, van Fraassen argues that a scientific explanation is a telling response to a why-question where such questions are identifiable by their *topics of concern, contrast classes*, and *explanatory relevance conditions*. An explanation is a *telling response* simply if it favors the occurrence of the state of affairs it is intended to explain. The *topic of concern* is the thing to be explained *(explanandum)*. The *contrast class* is the set of alternative possibilities, of which the topic of concern is a member, and for which an explanation might be requested in a particular context. The *explanatory relevance conditions* are the respects in which an answer might be given. For example, to borrow one of van Fraassen's illustrations, our topic of concern might be why an electrical conductor is warped. In this case, the contrast class might consist of other nearby conductors that are not warped, the warping of the conductor as opposed to its retaining its original shape, and so on. The explanatory relevance conditions might be the presence of a particularly strong magnetic field, the presence of moisture on the conductor, and so on. All of these things are highly dependent on context.

The pragmatic theory is relatively simple and direct in comparison with the other two models. It is also capable of accommodating the special aspects of the other two theories of explanation, and it has a very broad range of application. Critics of the pragmatic model have questioned whether every why-question asked by a scientist requires a contrast class, whether scientific questions might sometimes involve explanations of *how* as well as why (for example, the question of how genes replicate), whether a telling response must always favor the topic of concern, and whether the theory is too broad and would legitimize as scientific those explanations that the community of scientists might wish to exclude (though actual acceptance by the scientific community seems to be built into the criteria of legitimacy in this case).

In conclusion, there is no consensus among philosophers of science as to what constitutes a proper scientific explanation or what criteria a theory must possess in order to be truly scientific. *Despite extensive attempts, criteria that indisputably demarcate science from non-science or pseudo-science have never been offered.* The failure of these efforts gives us a strong reason to suspect that no such criteria exist. Even so, we will find it beneficial to pay another visit to these three models of scientific explanation when we address the question of whether theoretic-design explanations can be counted as scientific.

Overton Revisited

Returning to Overton's characterization of scientific theories, we find the requirement that they explain with reference to natural laws is not a universal characteristic of models for scientific explanation. Furthermore, the *testing intuition* that his account embodies, while an indispensable component, is not as clear in its application as the simplicity of its statement would suggest. Scientific theories are not usually testable without crucial auxiliary assumptions, and the question of what gets rejected (falsified) in the face of a failed experimental prediction is far from clear. We will see an example of this point when we discuss Smolin's cosmological theory in the next subsection.

I am no fan of creation science, but Overton's criteria do not straightforwardly establish the conclusion that it is unscientific. In claiming that the earth is no more than ten thousand years old, and that all fossilization is due to a universal flood of recent origin, the scientific creationist has made claims that are falsifiable. I would contend that these claims are not just falsifiable, but also have been

201

falsified. If so, then it would seem that the problem is not that creation science isn't a scientific theory, but rather that it is a bad one. It has been shown to be false and therefore should be rejected. Opponents can argue either that it is not a proper scientific theory, or that it is bad science, but they cannot make a reasonable case for both of these charges. Given the difficulties of specifying demarcation criteria that separate science from other human activities, they would be better off arguing that creation science should be rejected as bad science.

Of course, the question still remains whether the theory can be dismissed as unscientific because it does not refer to natural law, or, to put it more broadly, because it fails to abide by the principles of methodological naturalism. After all, creation science asserts that the earth's biological species did not evolve, but rather were the direct product of God's creative acts. If methodological naturalism is required for an explanation to be considered scientific, then invoking God's direct activity as an explanatory principle disqualifies creationism as science because it appeals to a cause that is not part of the physical universe. We need to take a closer look at the status of methodological naturalism as a constraint upon scientific theories and ask what science might look like if methodological naturalism were a context-dependent principle rather than a universal one.

Methodological Naturalism and the Limits of Science

There are many areas to which science cannot speak directly and many questions that it cannot answer. For instance, science cannot demonstrate the validity of its own methods, at least not without begging the question, nor can it be used to argue for the presuppositions that govern those methods. Science cannot speak directly to questions of the ultimate purpose of human existence, nor about the ultimate purpose of the universe itself. There certainly are scientists, Nobel laureate Steven Weinberg among them, who contend that the universe has no purpose. But such pronouncements do not by any means follow from the theories they construct, nor do they follow from any aspects of their scientific work. Science also cannot tell us what we ought to do, what is morally right and what is morally wrong. Finally, as we shall see in more detail shortly, science cannot provide an ultimate naturalistic explanation for the existence of the universe which it studies—there is a fundamental design problem that *will not* go away no matter how far back it is pushed. The question is whether this design problem might be brought into the scope of science itself, given that doing so might render methodological naturalism problematic in some contexts as a restriction on scientific theories.

Cosmology and Design

Let's look at a specific example of how fundamental design problems afflict naturalistic attempts at ultimate explanations. In cosmology, one question that confronts us is why certain natural quantities take on the values that they do. For instance, why do the mass and charge of an electron have the values that they do? Why do the forces of gravity and electricity have the strengths that they do? In short, why do all of the different constants of nature have the values that they have and not different ones? This question is made more compelling by the recent discovery that if the values of these constants had been different, sometimes even by astoundingly small amounts, the universe would not be stable and it would be impossible for life to exist.

The most intuitive explanation for this incredible string of coincidences is, of course, design: the universal constants were fine-tuned by an intelligent agent (or agents) to make the universe stable and hospitable for life. So far, the scientific community has rejected such a response as being outside the pale of science because it is interpreted as violating the canons of methodological naturalism. From the standpoint of pure logical possibility, it need not be interpreted this way. Perhaps our universe is embedded in another, much larger, physical universe and exists as the result of an experiment conducted by highly intelligent embodied beings who live in this larger universe. Of course, this only pushes the design problem back one step: where did *their* universe come from, and what are the conditions that made *it* possible? To avoid the specter of design altogether, a thoroughly naturalistic account of the origin of all possible physical universes, and of our own in particular, would have to be devised.

Cosmology and Natural Selection

One such naturalistic theory is theoretical physicist Lee Smolin's hypothesis of *cosmological natural selection*. Presented to a popular audience in his book *The Life of the Cosmos*,[3] his idea is fairly simple. Classical general relativity predicts the existence of black holes—singularities (points) in space-time where matter has shrunk to infinite density and the gravitational field has become infinitely strong. Quantum theory conflicts with this prediction because when physical objects shrink beyond a certain point in size, quantum effects take over and a system will no longer exhibit continuous causal behavior. One of the challenges in cosmology, therefore, is to devise an adequate quantum theory of gravity. Smolin conjectures that such a theory will preclude the notion of such singularities in space-time, and that what a black hole will turn out to be is a point at which space-time takes a "bounce" in another direction, creating a universe that is causally inaccessible from this one. In other words, black holes are actually universe generators.

Now this is pure speculation—it is not clear that a viable theory of quantum gravity will support Smolin's conjecture. All that Smolin knows is that this possibility is not precluded by present cosmological knowledge.

But Smolin goes even further: he proposes that at each bounce the values of the universal constants will change by a small random amount. Again, it is not known whether this is really possible, just that (for all we know right now) it might not be impossible. If it is possible, there still is no theoretical basis for the assumption that these changes will be either random or small. If they are not random, then they must conform to some pattern. If they conform to a pattern, then this suggests nomological (law-like) behavior. And if there is nomological behavior, the question of the origin of the laws presents itself. On the other hand, if the changes are random but not small, the hypothesis of cosmological natural selection will be a nonstarter, since it rests on an incremental evolution toward a stable range of values that maximize black hole production.

Assume that black holes are random universe generators (as in Smolin's model) and that certain values for universal constants are more conducive to black hole production than others. The cosmological natural selection hypothesis states that the natural evolution of cosmological systems will favor numerically those universes that produce the maximal number of black holes because they leave behind the most descendants. Assume also that our universe is a typical member of a collection of such universes. Then the hypothesis of cosmological natural selection explains the fine-tuned values of our universe's fundamental constants by suggesting that these values are precisely the ones that tend to maximize black hole production.

Smolin's theory is typical of the wildly speculative character of much of modern cosmology and the lengths to which many scientists will go to construct naturalistic explanations. It is clear in Smolin's book that his philosophical motivation involves more than just staying within the constraints of methodological naturalism. It involves a commitment to *metaphysical naturalism*, the doctrine that what is studied by the human and nonhuman sciences is all that there is. Thus any explanation that appeals to causes beyond the resources of the physical universe is necessarily false. Given Smolin's philosophical constraints, something like his theory *has* to be true, no matter how implausible it may seem, because there is no other option.

Testability

Smolin contends that the cosmological natural selection hypothesis is testable and potentially falsifiable because we can test the claim that the fundamental constants of our universe maximize black hole production. He outlines a number of procedures by which the theory might be tested. In this regard, consider the significance of the two possible outcomes of such tests. First, suppose his theory passes the tests. Given the wildly speculative character of his background assumptions, it still would be a stretch to say that we were warranted in believing that it is true. All we could conclude is that, for all we know, it is a *possible* explanation for why the fundamental constants of our universe have the values that they do.

But how far would this take us toward eliminating the design problem? Not very far at all, it would seem. Consider the fact that the vast majority of possible values for the fundamental constants of the universe would not permit the formation of black holes at all, let alone the existence of life. How is it that this process began in the neighborhood of parameter values that would allow a universe to generate at least one black hole (and hence at least one descendant)? And if the bounces were random, what kept it in this neighborhood? There is a design problem lurking here.

Or consider the proposed mechanism for generating universes. Smolin's black hole universe generators themselves must be governed by physical laws that permit the formation of universes. If these laws were different from what they are, presumably either the universe generator would not function and no universes would be produced, or all of the universes would be defective in a way that would preclude the stability of their existence. If there were no law of gravitation, for example, masses would not attract, and there would be no stars and planets, and hence no life. Or if the Pauli Exclusion Principle (which states that no two particles of matter can occupy the same quantum state) were not true, then electrons would collapse into the nucleus of the atom, and both atoms and the larger objects made of them would be impossible.

So it is not just the parameters of nature, but the laws too that must be explained. This is another design problem, and it cannot in principle be eliminated on a naturalistic basis. If we ask how the universe generator got going in the first place with the finely tuned laws that govern its function (even if the parameters in those laws fluctuate from universe to universe), then the metaphysical naturalist has no options other than to say that we were just incredibly lucky (which is no explanation at all) or to postulate another generator that generates universe generators. If he opts for the latter response, he is well on his way to an infinite regress of speculative explanations. In short, the design problem just *will not go away*.

On the other hand, suppose that Smolin's theory fails the tests he proposes. As indicated earlier, the falsifiability of a hypothesis depends on holding fixed the auxiliary assumptions needed to produce the required conclusion; otherwise the disconfirmation of the central hypothesis could be deflected onto one of these. In the case of cosmological natural selection, the central assumption to be tested was the claim that our universe is nearly optimal for black hole production. Note, however, that this claim about our universe is not actually a consequence of Smolin's theory itself, since the actual requirement is just that *most* universes are nearly optimal for black hole production. In order to get from this requirement to one that allows the theory to be tested in our universe, Smolin has to introduce the auxiliary assumption that our universe is typical. This auxiliary assumption, however, is neither testable nor required by the theory itself. This means that if we discovered that changes in the actual values of certain fundamental universal constants would substantially increase the number of black holes produced, we could still salvage Smolin's theory easily by denying the auxiliary assumption that our universe is typical. Of course, if we did this, cosmological natural selection would become a completely untestable hypothesis. Therefore, by Overton's criteria (and by the ones that Smolin himself proposes), it would fail as a scientific theory altogether. As a final blow, it also would cease to function as an explanation for the fine-tuning of the constants of *our* universe.

Given that methodological naturalism places any ultimate explanation for the origin of the universe beyond the scope of what science can offer, perhaps we should consider it to be an explanatory restriction appropriate in some contexts and undesirable in others. After all, we have just seen a particularly vivid example of how the background assumption of metaphysical and methodological naturalism led to an egregious exercise in "scientific" speculation. When we set this constraint aside, of course, we will still retain other intuitions such as the testability and fruitfulness of hypotheses, even though (as we have just seen) there can be problems with how such criteria are applied.

But there is little use in abstract speculation about what science *might* look like when methodological naturalism is suspended. Let's look at a concrete proposal, one which holds much promise as a research program.

Intelligent Design Theory

Recent work by a wide variety of scholars in many different disciplines has converged upon the notion that a rigorously articulated concept of design can be introduced fruitfully into the spectrum of possible scientific explanations. This involves the conviction that there can be well-defined criteria for determining when the concept applies. Let us start, therefore, by considering work done recently by the mathematician and complexity theorist William Dembski.

The Design Inference

Dembski draws attention to the fact that making design inferences is already an essential and uncontroversial part of various scientific activities ranging from the detection of fabricated experimental data, to forensic science, cryptography, and even the search for extraterrestrial intelligence (SETI). Citing an example from popular culture, Dembski notes how SETI researchers in the movie *Contact* classified a signal from an extraterrestrial source as being of intelligent origin on the basis of its reproduction of the sequence of prime numbers from 2 to 101. This mathematical sequence could not realistically have been generated by an unintelligent natural source. Rather, the signal exemplifies two trademarks that Dembski identifies as necessary and sufficient for inferring intelligence or design: complexity and specification.

Complexity ensures that the event in question is not so simple that it can readily be explained by chance. It is an essentially probabilistic concept. Specification ensures that the event in question exhibits the trademarks of intelligence. The notion of specification amounts to this: if, independently of the small probability of the event in question, we are somehow able to circumscribe and define it so as to render its reconstruction tractable, then we are justified in eliminating chance as the proper explanation for the event. Dembski calls such an event one of *specified small probability*. If an event of small probability fails to satisfy the specification criterion, it is still attributable to chance, as is the case, for example, with any sequence of heads and tails produced by one thousand tosses of a fair coin. But if an event is genuinely one of *specified* small probability, then the proper conclusion is that the cause of that event is intelligent agency. (For a more detailed explanation, see Dembski's "Signs of Intelligence," especially p. 180–88 of this volume.)

We can illustrate the notion of specified small probability with a couple of examples provided, ironically enough, by the evolutionary naturalist Richard Dawkins in his book *The Blind Watchmaker*. I offer them with some minor modifications. Suppose that a bank vault lock has a quadrillion possible combinations. Each of the quadrillion possible combinations is equally improbable, yet one of them in fact opens the lock. The actual combination that opens the vault is an event of specified small probability. If one person given one chance to open the vault succeeds in doing so, then the proper conclusion is that he opened the vault by design, namely by having prior knowledge of the right combination. Similarly, of all of the trillions of possible combinations of scrap metal, plastic, and rubber, very few can fly. One would not expect to hurl scrap metal around at random and wind up with a helicopter or an airplane. The arrangements that fly are capable of being specified in advance according to their function. Having a large collection of metal, plastic, and rubber in front of you that constitutes a Boeing 747 is an event of specified small probability. Even if you find it in a junkyard, the existence of the 747 cannot be attributed to the chance hurling around of spare parts by, say, a high wind. This intuitive understanding of specification is sufficient for this informal exposition. Dembski's important contribution has been to render the notion of specification mathematically rigorous in a way that places design inferences on a secure foundation.

Is Intelligent Design a Scientific Explanation?

Even though the statistical analysis used to determine whether an event is one of specified small probability is based on empirical observations, the design inference itself can be formulated as a valid deductive argument. One of its premises is a mathematical law that Dembski calls the *law of small probability*. That the design inference lends itself to this precision of expression seems quite significant. While the three models of scientific explanation we discussed earlier were all found to be inadequate, they did capture important intuitions, and we can easily see that the design inference meets the requirements of all of them.

The design inference conforms to the requirements of a deductive-nomological explanation because it satisfies all four criteria of this explanatory model.

1. The explanation it offers can be put in the form of a deductive argument.
2. It contains at least one general law (the law of small probability), and this law is required for the derivation of the thing to be explained (in this case the nature of the cause of the event in question).
3. It has empirical content because it depends on both the observation of the event and the empirical facts relevant to determining the objective probability of its occurrence.
4. The sentences constituting the explanation are true (to the best of our knowledge), because they take into account all of the relevant factors in principle available to us prior to the event we are seeking to explain.

It also satisfies the requirements of the causal-statistical model of explanation. The design inference procedure isolates the factors that are statistically relevant to the explanation of the event. It does so first by determining that the event in question is one of small probability; second, by ensuring that the criterion of specifiability is satisfied; and last, by therefore eliminating natural law and chance as possible explanations. It also satisfies the requirement of making manifest the causal network undergirding the statistical regularities, since it causally connects the relevant explanatory factor (intelligent agency) to the occurrence of the event.

Finally, it satisfies the pragmatic model of explanation because it provides a telling answer to a why-question, where that question is identifiable by its topic of concern, its contrast class, and its explanatory relevance conditions. The topic of concern is the observed occurrence of an improbable event that bears *prima facie* evidence of specification. The contrast class is constituted by the set of alternatives of which the topic of concern is a member. For example, the contrast class might include the occurrence of other more probable events in the causal context under consideration, or equally improbable events in that context that bear no evidence of specification, etc. The explanatory relevance conditions might be the presence of highly particular initial conditions in the physical system, indications of thermodynamic counter-flow, the presence of apparently intelligent informational content, and so on. All of these things are dependent upon the context, but what is sought is a correct account of the cause of the event in question. The response provided by the design inference is therefore a telling one by the standards of the pragmatic model, because when an event of demonstrably specified small probability occurs, this state of affairs is favored by the design-theoretic explanation.

Thus the design inference satisfies all three models of scientific explanation, and there seems little reason to bar its legitimacy as a mode of scientific explanation. Indeed, when generating scientific conclusions in cryptography or forensics, the design inference is not controversial. The sticking point is with the philosophical issue of methodological naturalism. What happens if the design inference, applied to certain natural phenomena, yields the conclusion that there is an intelligent cause that might transcend our universe? There seems to be an illegitimate double standard operative in barring such a conclusion when design inferences are otherwise scientifically acceptable.

Individual conclusions are one thing, but the possibility of a general science of intelligent design is another. Most critics have objected

210

that intelligent design, if admitted as a legitimate explanation in the natural sciences, would halt scientific progress by curtailing inquiry. After all, if we explain an event as the result of the direct activity of an intelligent agent or agents that are not a part of our spatio-temporal framework, what more can be said?

Can Intelligent Design Theory Generate a Research Program?

Is intelligent design (ID) theory a "science-stopper"? In a word, no. Every viable research program generates its own set of questions. While many of the questions that are natural to an ID approach to science will be different from those that are pursued under the rubric of evolutionary naturalism, they will be no less scientific and, I suspect, no less fruitful. It is also important to realize that ID theorists are *not* advocating a cessation of research from the perspectives of neo-Darwinian or self-organizational complexity theories. These approaches are tremendously interesting and fruitful in their own right, and it would be foolish to abandon them. Since science proceeds by a largely self-corrective process, having all three models interacting would highlight the strengths and the weaknesses of each, as well as the challenges and limitations peculiar to different explanatory strategies. These three approaches have things to learn from each other, and it seems likely that a clearer picture of what is happening in nature will not emerge from one model in isolation, but from the confluence of contributions from all three.

The Idea of Irreducible Complexity

One of the central concepts in ID theory is that of *irreducible complexity*. Since it is the focus of a number of research concerns, we need to have an informal definition of it in hand. We say that a system is irreducibly complex if it consists of several interrelated parts such that the removal of any one of these parts completely destroys the system's function. A simple example of an irreducibly complex system is a mousetrap consisting of a platform, hammer, spring, catch, and holding bar (see pp. 93–94 of this volume). If any one of these five components is removed, the mousetrap is rendered nonfunctional.

211

Contrasted with irreducible complexity is what might be called *cumulative complexity*. A system is cumulatively complex if its components can be arranged sequentially in such a manner that successive removal never results in complete loss of function. The neo-Darwinian model of selection supervening on random genetic mutation requires cumulative complexity, as does self-organizational complexity theory, which attempts to ratchet increased order out of systems disturbed from their equilibrium states. Organic evolution takes place in both of these models by the accretion of beneficial components that allow life to move from states of lesser to greater complexity.

If irreducibly complex systems exist, however, neither a random neo-Darwinian nor a self-organizational mechanism will be able to produce them through selection effects. If selective mechanisms acted with reference to a prespecified goal, then irreducibly complex systems might result; but since selection relies on undirected natural processes (ones that have no end in view) they cannot. So both neo-Darwinian and self-organizational theorists, if their programs are taken as comprehensive research strategies, must maintain that *all* complexity in nature is cumulative complexity. The reason for this is that in an irreducibly complex system, function is only attained when *all* of the components are in their respective places. If any part of an irreducibly complex system is missing, the system is completely nonfunctional. Since natural selection can only choose systems that are already working, it cannot generate such a system by cumulative means. The only way to produce such a system is all at once or not at all. If all such systems had very few components this might be possible, but the problem is they do not.

Consider any irreducibly complex system. The integrated complexity of this system constitutes a measure of its improbability, and if the system is sufficiently complicated its existence will constitute a small probability event. Furthermore, the functionality of the system will represent a pattern that satisfies the two specifiability conditions of the design inference: it is detachable from the probability of the pattern coming into existence by chance, and it is tractable in the sense of being constructible on the basis of background knowledge of the function it performs. This, at least, is the theoretical picture.

There are many details that need to be worked out. We must be able to map a mathematically rigorous definition of irreducible complexity (constructed using the tools of probability, complexity, information, stochastic process, and recursion theory) onto cosmological, physical, genetic, biochemical, and biomechanical systems in a realistic and unambiguous way if the concept is going to have any empirical bite. Some preliminary work has been done on the definitional problem by mathematicians David Berlinski and William Dembski. Some applicational studies have been undertaken by biochemist Michael Behe, biologists Paul Nelson, Jonathan Wells, and Siegfried Scherer, physical anthropologist Sigrid Hartwig-Scherer, philosopher of biology Stephen Meyer, and philosophers of physics Robin Collins and me—but there still is much work to do.

The Genesis of a Research Program

The research program associated with design, broadly conceived, can be divided into three major areas: the scientific determination of design, the scientific outworking of design, and the philosophical and social implications of design. With respect to the determination of design, as indicated above, there is work that remains to be done on its precise mathematical definition. This will involve:

1. the further articulation and honing of Dembski's probability and complexity theoretic model and the development of its information theoretic counterpart;
2. the construction of a precise definition of irreducible complexity using, among other things, the mathematical resources associated with stochastic process theory, probability theory, mathematical logic, recursion theory, computer science, and the theory of cellular automata; and
3. the articulation of a quantificational measure of the information in physical systems, the search for information-theoretic conservation laws in this context, and an analysis of known physical laws from this perspective, perhaps using models developed on the basis of Fisher information (see, e.g., B. Roy Frieden's *Physics from Fisher Information: A Unification*, Cambridge University Press, 1998).

These resources then can be applied to an analysis of cosmological and biological structures for the purpose of detecting design, and such attempts at application undoubtedly will lead to further refinements.

With regard to the detection of design, there is no area of physical or biological science that is not a potential area of investigation: the structure of physical laws, the fine-tuning of fundamental physical constants, and the organization of cosmological, astronomical, astrophysical, solar, planetary, geological, meteorological, ecological, zoological, paleontological, biological, genetic, biochemical, molecular biological, organic and inorganic chemical, quantum chemical, and quantum mechanical systems are all ripe for analysis from a design-theoretic perspective. Potentially, there is also room for its application to an analysis of cognitive psychological, linguistic, sociological, political, and economic systems, questions of historical development, and even an analysis of the effectiveness of mathematical and statistical description in the physical sciences.

If design is detected in a physical or biological system, its scientific outworking will yield a variety of natural research possibilities, such as:

1. the details of the construction of the system as a problem in reverse engineering;
2. the determination of the proper functions and purposes of the system;
3. an analysis from the perspective of signal engineering to determine how noise, age, friction, and/or mutation might have obscured the original design plan and contributed to dysteleology;
4. the reconstruction of the original design plan;
5. the limits of variability in design, that is, the constraints within which the system functions well, and outside of which it malfunctions or ceases to function altogether (e.g., genetic knockout experiments, etc.);
6. the developmental capacity of the system—how far it can increase in complexity without intelligent informational influx;
7. the optimality of the design subject to the multiple constraints imposed by the different functions and purposes of the system;
8. the dynamics of interactions between systems, and the developmental possibilities and constraints inherent in such interactions;
9. the integration and mutual support relations among different but related complex systems; and

214

10. the technological application of all of the foregoing to concerns in (among other things) medicine, environmental science, and systems theory.

If such research is successful, some of the consequences will be not only a deeper theoretical understanding of various structures and processes in nature (and perhaps the discovery of new ones), but also practical benefits in the medical treatment of disease, strategies for dealing with the environmental problems plaguing first-world cultures, and advances in information technology. This is not to say, of course, that progress on these things is not possible from any other research perspective, just that the design-theoretic perspective and its accompanying tools may reveal some things about the constitution and function of nature that would not be evident or readily emerge from a different angle of approach.

Further Implications

Finally, beyond the realm of strict scientific concern, there are philosophical and social implications of design. The design-theoretic perspective has conceptual ramifications that extend not just through the nature and practice of science as we have discussed, but through general questions of ontology, epistemology, ethics, and public policy as well.

Natural questions arise as to the identity of the intelligence behind any design of cosmological and biological structures, and our identity and place as human beings in such an order. From the standpoint of the inherent limitations on human cognitive capacity and the process of knowledge acquisition, a question can arise about the epistemological benefits that would ensue from having cognitive systems that are intelligently designed. If our cognitive systems arose as the result of undirected natural processes, it is unlikely that their operation would be conducive to the discovery of truth, especially in theoretical disciplines that have little direct bearing on survival. On the other hand, if our cognitive systems are designed, a different and more optimistic conclusion might follow.[4]

Design theory also leads readily to an examination of whether there are natural ethical principles that work as constraints within which both individuals and society are designed to function, and

215

outside of which personal and societal malfunction of various degrees would be expected. The tools of anthropological and sociological study may be applied to this question, and some sort of empirical-statistical analysis should be possible. Finally, there is the application of the results of all of the foregoing investigations to the formulation of public policy objectives aimed at helping our society, culture, and government to function more smoothly.

Conclusion

I hope that a number of crucial things have been made clear to the reader in this discussion. First, since science and the scientific community play such a large role in modern culture, it is important to spend some time in concentrated reflection on the philosophical presuppositions that have shaped the methodology and content of modern science. Second, while science is eminently worthy of our respect, the grounding of its theories and conclusions, as well as the objectivity of its methodology, are not as unproblematic as popular conception would have us believe. Third, methodological naturalism has been a pervasive constraint on scientific theorizing and practice, and it artificially restricts the range of acceptable theoretical options. Fourth, from the consideration of various accounts of scientific explanations and their limitations, it is clear that design-theoretic inferences meet the standard "requirements" of scientific explanations. Only the constraint of methodological naturalism stands in the way of applying this tool to cosmological, physical, and biological structures, and, as I have argued, it is possible to do science in certain contexts without this constraint. Finally, design theory can generate a robust research program capable of supplementing the techniques of neo-Darwinian evolutionary biology and self-organizational complexity theory in a rigorous and significant way. By so doing, it offers the possibility of solutions to various difficulties that so far have proven intractable to these other approaches.

NOTES

Introduction What Intelligent Design Is Not

1. Henry Petroski, *Invention by Design: How Engineers Get from Thought to Thing* (Cambridge, Mass.: Harvard University Press, 1996), p. 30. Petroski is a professor of civil engineering as well as a professor of history at Duke University.

2. For a critique of Gould's objections to design based on optimality see Paul Nelson, "The Role of Theology in Current Evolutionary Reasoning," *Biology and Philosophy* 11, (1996): 493–517.

3. Consider, for instance, the inverted design of the vertebrate retina, seized on by generations of Darwinists as a maladaptation for placing photoreceptors behind rather than in front of neural connectors and thus entailing a blind spot. It now appears that "the very high energy demands of the photoreceptor cells in the vertebrate retina suggest that rather than being a challenge to teleology the inverted design of the vertebrate retina may in fact represent a unique solution to the problem of providing the highly active photoreceptor cells of higher vertebrates with copious quantities of oxygen and nutrients." Michael Denton, "The Inverted Retina: Maladaptation or Pre-adaptation?" *Origins & Design* 19(2), 1999: 15.

4. Francis Darwin, ed., *The Life and Letters of Charles Darwin*, vol. II (New York: D. Appleton and Co., 1888), p. 105.

5. Charles Darwin, *On the Origin of Species*, facsimile 1st ed. (Cambridge, Mass.: Harvard University Press, 1964 [1859]), pp. 242–244.

6. Stephen Jay Gould, *The Panda's Thumb* (New York: Norton, 1980), pp. 20–21.

7. See Boethius, *The Consolation of Philosophy*, in *Loeb Classical Library* (Cambridge, Mass.: Harvard University Press, 1973), p. 153. Alvin Plantinga's free will defense is a resolution of the problem of evil that has provoked much response from philosophers of religion—for a synopsis see Kelly James Clark, *Return to Reason* (Grand Rapids, Mich.: Eerdmans, 1990), ch. 2. Finally, a significant number of contemporary philosophers of religion resolve the problem of evil by denying traditional accounts of divine omniscience and omnipotence. Process theologians have taken this view for some time, but more traditional philosophers and theologians are now taking this line also—see William Hasker, *God, Time, and Knowledge* (Ithaca, N.Y.: Cornell University Press, 1989).

8. See his review of Michael Behe's *Darwin's Black Box* in James A. Shapiro, "In the Details . . . What?" *National Review*, September 19, 1996: 62–65.

9. John Haught, for instance, thinks intelligent design is thoroughly objectionable theologically. See his *God After Darwin: A Theology of Evolution* (Boulder, Colo.: Westview Press, 2000).

10. Cf. Larry Arnhart in, "Conservatives, Darwin & Design: An Exchange," *First Things*, November 2000, pp. 23–31.

11. Richard Feynman, *"Surely You're Joking, Mr. Feynman!"* (New York: Bantam, 1986), p. 313.

12. Richard Dawkins, review of Donald Johanson and Maitland Edey's *Blueprints*, *New York Times*, April 9, 1989, sec. VII, p. 34.

13. Michael Ruse, *Darwinism Defended* (Reading, Mass.: Addison-Wesley, 1982), p. 58.

14. Michael Shermer, *Why People Believe Weird Things* (New York: W. H. Freeman, 1997), p. 148.

15. See Ronald Numbers, *Darwinism Comes to America* (Cambridge, Mass.: Harvard University Press, 1998), pp. 9, 11.

16. Daniel Dennett, *Darwin's Dangerous Idea* (New York: Simon & Schuster, 1995), p. 519.

17. Robert Pennock, *Tower of Babel: The Evidence Against the New Creationism* (Cambridge, Mass.: MIT Press, 1999).

18. See also Kenneth Miller, *Finding Darwin's God* (New York: Harper Collins, 1999).

19. Pennock, *Tower of Babel*, p. 295.

20. This is made abundantly clear in F. H. Sandbach, *The Stoics*, 2d ed. (Indianapolis: Hackett, 1989), especially ch. 4.

21. Richard Dawkins, *The Blind Watchmaker* (New York: W. W. Norton & Co., 1987), pp. 85–86.

22. See respectively Alan Guth, *The Inflationary Universe* (Reading, Mass.: Addison-Wesley, 1997); Lee Smolin, *The Life of the Cosmos* (New York: Oxford University Press, 1997); Peter Atkins, *Creation Revisited* (Harmondsworth: Penguin, 1994).

23. See respectively Jacques Monod, *Chance and Necessity* (New York: Vintage, 1972); Dawkins, *The Blind Watchmaker;* Stuart Kauffman, *At Home in the Universe* (New York: Oxford University Press, 1995).

24. Quoted in Moshe Sipper and Edmund Ronald, "A New Species of Hardware," *IEEE Spectrum* 37(4), April 2000: 59.

25. See Sandbach, *The Stoics*, pp. 14–15. Also, see Ben Wiker's forthcoming *The Christian and the Epicurean*, InterVarsity Press.

26. Richard Swinburne, *The Existence of God* (Oxford: Clarendon, 1979), ch. 8, "Teleological Arguments."

27. Paul Davies, *The Mind of God* (New York: Touchstone, 1992), ch. 8, "Designer Universe."

28. Richard Lewontin, "Billions and Billions of Demons," review of *The Demon-Haunted World: Science as a Candle in the Dark* by Carl Sagan, *New York Review of Books*, January 9, 1997: 31.

29. Consult the index of any popular book on complex self-organization, for instance, Peter Coveney and Roger Highfield, *Frontiers of Complexity* (New York: Fawcett Columbine, 1995).

30. Dawkins, *The Blind Watchmaker*, p. 1.

Chapter 1 The Intelligent Design Movement

1. p. viii-ix

2. p. 4

3. Judith Hooper, "A New Germ Theory," *Atlantic Monthly*, 283 (2): 41-53.

4. Jonathan Weiner, *The Beak of the Finch* (New York: Knopf, 1994).

5. Jerry A. Coyne, "Not Black and White," *Nature* 396 (1998): 35-6.

6. Edward Wilson, *Consilience* (New York: Knopf, 1998).

7. Michael Behe, *Darwin's Black Box: The Biochemical Challenge to Evolution* (New York: Free Press, 1996).

8. William A. Dembski, *The Design Inference: Eliminating Chance through Small Probablilities* (Cambridge: Cambridge University Press, 1998).

Chapter 2 Design and the Discriminating Public

1. Percival Davis and Dean H. Kenyon, *Of Pandas and People: The Central Question of Biological Origins* (Dallas, Texas: Haughton Publishing Co., 1993).

2. Wheaton, Ill.: Crossway Books, 1994.

3. Wheaton, Ill.: Tyndale House Publishers, 1999.

4. Phillip Johnson, *Reason in the Balance: The Case against Naturalism in Science, Law, and Education,* (Downers Grove, Ill.: InterVarsity Press, 1998).

5. Francisco J. Ayala, "Darwin's Revolution," in John H. Campbell and J. William Schopf, eds., *Creative Evolution!?* (Boston: Jones & Bartlett Publishers, 1994), p. 5.

6. Daniel C. Dennett, *Darwin's Dangerous Idea: Evolution and the Meaning of Life* (New York: Simon & Schuster, 1995), p. 63.

7. Richard Dawkins, *The Blind Watchmaker: Why the Evidence of Evolution Reveals a Universe Without Design* (New York: W. W. Norton & Co., 1996), p. 6.

8. Cited in Gertrude Himmelfarb, *Darwin and the Darwinian Revolution* (Garden City, NY: Doubleday Anchor Books, 1959), pp. 329-30.

9. Jacques Barzun, *Darwin, Marx, Wagner: Critique of a Heritage,* 2d ed. (Chicago: University of Chicago Press, 1981), pp. 11, 36.

10. The video is titled "Darwinism: Science or Naturalistic Philosophy?" and is available from Access Research Network, http://www.arn.org.

11. Johnson, *Reason in the Balance*, pp. 46-7.

12. Walter L. Bradley, Charles Thaxton, and Roger L. Olsen, *The Mystery of Life's Origin* (Dallas: Lewis and Stanley, 1993).

13. Nancy R. Pearcey "The Evolution Backlash: Debunking Darwin," *World* 11 (38) March 1, 1997: 12–15.

Chapter 3 Proud Obstacles and a Reasonable Hope

1. Jay Wesley Richards and William A. Dembski, *Unapologetic Apologetics: Meeting the Challenges of Theological Studies* (Downers Grove, Ill.: InterVarsity Press, 2001).

2. Richard Lewontin, "Billions and Billions of Demons."

3. J. Gresham Machen, "Christianity and Culture," in *What Is Christianity? And Other Addresses,* ed. Ned Stonehouse (Grand Rapids: Eerdmans, 1951), p. 162.

Chapter 4 The Regeneration of Science and Culture

1. John G. West Jr., *The Politics of Revelation and Reason* (St. Lawrence: University Press of Kansas, 1996).

2. John G. West Jr., Jeffrey P. Schultz, Ian MacLean, George Kurian, eds., *The Encyclopedia of Religion in American Politics* (Phoenix: Oryx Press, 1999).

3. Jack London, *The Call of the Wild* (New York: MacMillan Co., 1903).

4. Washington Gladden, *Present Day Theology*, 3d ed. (Columbus: McClelland and Company, 1913), pp. 36-7.

5. Marvin Olasky, *The Tragedy of American Compassion* (Washington, D.C.: Regnery Publishing, 1995).

6. Ludwig Büchner, *Force and Matter*, 4th English ed. (New York: Peter Eckler, Publisher, 1891).

7. Ibid, p. 376.

8. Ibid, p. 378.

9. Nathaniel F. Cantor, *Crime, Criminals and Criminal Justice* (New York: Henry Hold and Company, 1932), p. 265–66.

10. Peter Singer, "Sanctity of Life or Quality of Life?" *Pediatrics* (July 1983): 128-9.

11. Robert Wright, *The Moral Animal* (New York: Random House, 1995).

12. Richard Dawkins, *Unweaving the Rainbow* (Boston: Houghton Mifflin, 1998).

13. John O. McGinnis, "The Origin of Conservatism," *National Review* (December 22, 1997): 31.

14. James Q. Wilson and Larry Arnhart, *Darwinian Natural Right: The Biological Ethics of Human Nature* (Albany: State University of New York Press, 1998).

15. C. S. Lewis, *The Abolition of Man* (New York: MacMillan Publishing Company, 1955), p. 90.

Chapter 5 The World as Text

1. William A. Dembski, ed., *Mere Creation: Science, Faith and Intelligent Design* (Downers Grove, Ill.: InterVarsity Press, 1998).

2. David C. Lindbert, *The Beginnings of Western Science* (Chicago: University of Chicago Press, 1993).

3. Lynn White, *Medieval Technology and Social Change* (Oxford: Clarendon Press, 1966).

4. Henri Cardinal de Lubac, *Medieval Exegesis,* trans. Mark Seganc (Grand Rapids: Wm. B. Eerdmans, 1998).

5. Ibid, p. 76f.

6. Roger Lundin, *Emily Dickinson and the Art of Belief* (Grand Rapids: Wm. B. Eerdmans, 1998).

Chapter 6 Getting God a Pass

1. John Mark Reynolds and J. P. Moreland, eds., *Three Views on Creation and Evolution* (Grand Rapids: Zondervan, 1999).

2. J. B. Morris and W. H. Simcox, trans., *The Homilies of St. John Chrysostom on the Epistle of St. Paul the Apostle to the Romans*, revised by George B. Stevens in *Nicene and Post-Nicene Fathers*, First Series, ed. P. Schaff (Peabody, Mass.: Hendrickson, 1994), p. 352.

3. J. P. Moreland, *Christianity and the Nature of Science* (Grand Rapids: Baker Books, 1989).

4. Phillip Johnson, *Darwin on Trial* (Washington, D.C.: Regnery Publishing, Inc., 1991).

Chapter 7 Darwin's Breakdown

1. *Science* 277(1997): 892.

2. Joseph Ratzinger, *In the Beginning: A Catholic Understanding of the Story of Creation and the Fall* (Grand Rapids: Wm. B. Eerdmans, 1986).

3. Ibid, p. 54-6.

4. Charles Darwin, *Origin of Species,* 6th ed. (New York: New York University Press, 1988), p. 154.

5. James Shreeve, "Design for Living," *New York Times,* August 4, 1996, sec. 7, p. 8.

6. J. A. Shapiro, "In the Details . . . What?" *National Review,* September 16, 1996, 62-5.

7. J. A. Coyne, "God in the Details," *Nature* 383 (1996): 227–8.

8. A. Pomiankowski, "The God of the Tiny Gaps," *New Scientist,* September 14, 1996: 44-5.

Chapter 8 Word Games

1. Gary B. Ferngren, Edward J. Larson, Darrel W. Amundsen, eds., *The History of Science and Religion in the Western Tradition* (New York: Garland Publishing, 2000).

2. Jon Buell and Virginia Hearn, eds., *Darwinism: Science or Philosophy* (Richardson, Tex.: Foundation for Thought and Ethics, 1994).

3. Percival Davis and Dean H. Kenyon, *Of Pandas and People.*

4. J. P. Moreland, ed., *The Creation Hypothesis* (Downers Grove, Ill.: InterVarsity Press, 1994).

5. Jitse M. van der Meer, ed., *Facets of Faith and Science* (Lanham, Md.: University of America Press, Inc., 1996).

6. *Cell,* 92(3).

7. A. G. Cairns-Smith, *The Life Puzzle* (Edingurgh: Oliver and Boyd, 1971), p. 95.

8. Dean H. Kenyon and Gary Steinman, *Biochemical Predestination* (New York: McGraw-Hill Book Co., 1969).

Chapter 9 Making Sense of Biology

1. Jonathan Wells, *Charles Hodge's Critique of Darwinism* (Caredigion, Wales: Edwin Mellen Press, 1988).

2. Jonathan Wells, *Icons of Evolution* (Washington: Regnery Publishing, 2000).

3. William Paley, *Natural Theology* (London: Wilks and Taylor, 1802).

4. Richard Dawkins, *The Blind Watchmaker,* p. 287.

5. Theodosius Dobzhansky, "Nothing in Biology Makes Sense Except in the Light of Evolution," *The American Biology Teacher,* 35 (1973): 125-9.

Chapter 10 Unfit for Survival

1. Paul A. Nelson, "The Role of Theology in Current Evolutionary Reasoning," *Biology and Philosophy* 11 (1996): 493–517.

2. Jacques Monod, *Chance and Necessity* (New York: Vintage Books, 1971).

3. Ibid, p. 7.

4. Ibid, p. 17.

5. Charles Darwin, *On the Origin of Species* (London: Harvard University Press, 1964), p. 3.

6. Richard Lewontin, "Adaptation," *Scientific American* 239: 212-30.

7. Darwin, *Origin of Species,* pp. 4-5.

8. J. Hodge, "The Development of Darwin's General Biological Theorizing," in *Evolution from Molecules to Men,* ed. D. S. Bendal (Cambridge: Cambridge University Press, 1983), p. 45.

9. Darwin, *On the Origin of Species,* p. 48.

10. Ibid, p. 61.

11. R. Dunbar, "Adaptation, Fitness, and the Evolutionary Tautology," *Current Problems in Sociobiology*, ed. King's College Sociobiology Group (Cambridge: Cambridge University Press, 1982), p. 10.

12. Ronald Brady, "Natural Selection and the Criteria by Which a Theory Is Judged," *Systematic Zoology* 28: 600-21.

13. Ronald Brady, "Dogma and Doubt," *Biological Journal of the Linnean Society* 17: 79-96.

14. J. G. Ollason, "What Is This Stuff Called Fitness?" *Biology and Philosophy* 6: 81-92.

15. Lewontin, "Adaptation," p. 122.

16. Peter Saunders and M. W. Ho, "Is Neo-Darwinism Falsifiable—And Does It Matter?" *Nature and System* 4 (1982): 179-196.

17. Brady, "Natural Selection," p. 606.

18. Saunders and Ho, "Is Neo-Darwinism Falsifiable," p. 182.

19. Ollason, "What Is This Stuff Called Fitness?" p. 91.

20. Leigh Van Valen, "Three Paradigms of Evolution," *Evolutionary Theory* 9: 1-17.

21. Elliott Sober, *The Nature of Selection* (Cambridge, Mass.: MIT Press, 1984), p. 61.

22. Stephen Stearns and Paul Schmid-Hempel, "Evolutionary Insights Should Not Be Wasted," *Oikos* 49: 118-125.

23. Bruce Naylor and Paul Handford, "In Defense of Darwin's Theory," *BioScience* 35: 473-484.

24. Donn Rosen, "Darwin's Demon," *Systematic Zoology* 27: 370-373.

25. Joel Cracraft, "The Use of Functional and Adaptive Criteria in Phylogenetic Systematics," *American Zoologist* 21: 21-36.

26. Sober, *Nature of Selection*, p. 62.

27. Darwin, *On the Origin of Species*, p. 6.

28. George Williams, *Adaptation and Natural Selection* (Princeton: Princeton University Press, 1966), p. 251.

29. T. Dobzhansky, F. Ayala, G. Stebbins, and J. Valentine, *Evolution* (San Francisco: W. H. Freeman, 1977), p. 504.

30. John Maynard Smith, *The Theory of Evolution* (New York: Penguin), 1975.

31. Richard Dawkins, *The Extended Phenotype* (San Francisco: W. H. Freeman, 1982), p. 19.

32. Ernst Mayr, Foreword to M. Ruse, *Darwinism Defended* (Reading, Mass.: Addison-Wesley, 1982), p. xi-xii.

33. John Endler, *Natural Selection in the Wild* (Princeton: Princeton University Press, 1986), pp. 46, 248.

34. Ibid, p. 3.

35. Ibid, p. 4.

36. Ibid, p. 46.

37. Richard Dawkins, "Replicators and Vehicles," in *Current Problems in Sociobiology,* ed. Kings College Sociobiology Group (Cambridge Univ. Press, 1982), p. 45.

38. C. Gans and R. Northcutt, "Neural Crest and the Origin of Vertebrates: A New Head," *Science* 220: 268-274.

39. Ibid, p. 272.

40. Darwin, *On the Origin of Species*, p. 108.

41. Endler, *Natural Selection in the Wild*, p. 248.

42. Soren Løvtrup, "Semantics, Logic, and Vulgate Neo-Darwinism," *Evolutionary Theory* 4: 157-172.

43. Michael Bradie and Mark Gromko, "The Status of the Principle of Natural Selection," *Nature and System* 3: 3-12.

44. Arthur Caplan, "Say It Just Ain't So: Adaptational Stories and Sociobiological Explanations of Social Behavior," *Philosophical Forum* 13: 144-160.

45. Ibid, pp. 149-150.

46. Gerhard Müller, "Experimental Strategies in Evolutionary Embryology," *American Zoologist* 31: 605-615.

47. Løvtrup, "Semantics, Logic, and Vulgate Neo-Darwinism," p. 178.

Chapter 11 The Cambrian Explosion

1. John L. Wiester, *The Genesis Connection,* 2d ed. (Nashville: Thomas Nelson Publishers, 1983).

2. Committee for Integrity in Science Education Staff, *Teaching Science in a Climate of Controversy* (Ipswich, Mass.: American Science Affiliation, 1989).

3. John L. Wiester and Robert C. Newman, *What's Darwin Got to Do with It?* (Downers Grove, Ill.: InterVarsity Press, 2000).

4. Quoted in *Creative Evolution!?*, ed. J. H. Campbell and J. W. Schopf (Sudburg, Mass.: Jones and Bartlett, 1994) p. 4-5.

5. *The Triumph of Evolution and the Failure of Creationism,* (New York: W. H. Freeman & Co., 2000); p. 42.

6. Thomas Henry Clark and Colin William Stearn, *Geological Evolution of North America* (New York: Ronald Press Co., 1960).

7. Ibid., p. 43.

Chapter 12 The "Just So" Universe

1. William Paley, *Natural Theology* (London: Wilks and Taylor, 1802).

2. Johannes Kepler, *Defundamentis Astrologiae Certioribus,* Thesis XX (1601).

3. Morris Kline, *Mathematics: The Loss of Certainty* (New York: Oxford University Press, 1980).

4. Ibid., p. 52.

5. Eugene Wigner, "The Unreasonable Effectiveness of Mathematics in the Physical Sciences," *Communications on Pure and Applied Mathematics* 13 (1960): 1-14.

6. Albert Einstein, *Letters to Solovine* (New York: Philosophical Library, 1987), p. 131.

7. Richard Courant, *Partial Differential Equations,* vol. II of R. Courant and D. Hilbert, *Methods of Mathematical Physics* (New York: Interscience Publishers, 1962) pp. 765–766.

8. John Barrow and Frank Tipler, *The Anthropic Cosmological Principle* (Oxford: Clarendon Press, 1988).

9. John Leslie, *Universes* (New York: Routledge, 1989).

10. Paul Davies, *The Accidental Universe* (Cambridge: Cambridge University Press, 1982).

11. Paul Davies, *Superforce* (Portsmouth, N.H.: Heinemann, 1984).

12. Paul Davies, *The Cosmic Blueprint* (Portsmouth, N.H.: Heinemann, 1988).

13. John Gribbin and Martin Rees, *Cosmic Coincidences* (New York: Bantam Books, 1989).

14. Reinhard Breuer, *The Anthropic Principle*, trans. Harry Newman and Mark Lowery (Boston: Birkhäuser, 1991).

15. Gilles Cohen-Tannoudji, *Universal Constants in Physics*, trans. Patricia Thickstun (New York: McGraw-Hill, 1993).

16. J. P. Moreland, ed., *The Creation Hypothesis* (Downers Grove, Ill.: InterVarsity Press, 1994)

17. Lawrence M. Krauss, "Cosmological Antigravity," *Scientific American,* January 1999: 53-59.

Chapter 13 Signs of Intelligence

1. William A. Dembski, *Intelligent Design: The Bridge Between Science and Theology* (Downers Grove, Ill.: InterVarsity Press, 1999).

2. Jacques Monod, *Chance and Necessity* (New York: Knopf, 1971).

3. Eliot Marshall, "Medline Searches Turn Up Cases of Plagiarism," *Science* 279 (January 1998): 473-74.

4. Charles Darwin, *On the Origin of Species* (Cambridge, Mass.: Harvard University Press, 1963), p. 482.

Chapter 14 Is Intelligent Design Science?

1. G. Brittan and K. Lambert, *Philosophy of Science* (Montreal: McGill-Queen's University Press, 1992)

2. M. Salmon, et al., *Introduction to the Philosophy of Science* (Upper Saddle River: Prentice-Hall, Inc., 1992).

3. Lee Smolin, *The Life of the Cosmos* (Oxford: Oxford University Press, 1997).

4. For an extended discussion of these ideas see Alvin Plantinga, *Warrant and Proper Function* (Oxford: Oxford University Press, 1993) and *Warranted Christian Belief* (Oxford: Oxford University Press, 2000).